Charles Daubeny

Essay on the Trees and Shrubs of the Ancients

Charles Daubeny

Essay on the Trees and Shrubs of the Ancients

ISBN/EAN: 9783337225872

Printed in Europe, USA, Canada, Australia, Japan

Cover: Foto ©berggeist007 / pixelio.de

More available books at **www.hansebooks.com**

. ON THE

RUBS OF T

THE SUBSTA

ur Lectu

DELIVERED BEFOR

ERSITY O

E SUPPLEMENTARY

N HUSBAN

READY PUBLISHED

BY

NY, M.D

AND RUR

N HENRY AND JAMES

1865.

PREFACE.

THE late Professor Sibthorp, in founding a Chair of Rural Economy to be attached to that of Botany already existing in the University of Oxford, directed that the Holder of that Office should deliver each Term a Lecture on some one of the subjects which the Professorship in question might be regarded as embracing.

In conformity to this regulation, I have, besides placing before my hearers from time to time the newest views on the theory of Agriculture which modern science had developed, given occasionally such sketches of the Husbandry of the Ancients, as could be gathered from the *Scriptores Rei Rusticæ* whose writings have come down to us.

The latter have since been embodied in a work published by me in 1857, entitled " Lectures on Roman Husbandry," in which I presented " an account of the System of Agriculture, of the Treatment of Domestic Animals, and of the Horticulture pursued in Ancient Times," concluding with notices " of the Plants mentioned in Columella and Virgil."

To this Publication the present Lectures may be

regarded as supplementary, containing, as they do,
a summary of the best information I have been able
to collect as to the Trees and Shrubs really intended
by those described or noticed in the principal Greek
and Roman writers.

For this purpose I have availed myself, not only
of the researches of Dr. Sibthorp himself, and of the
old commentaries on Theophrastus and Dioscorides,
which were accessible at the time he left the materials for his *Prodromus Floræ Græcæ*, but also of
various works on the Flora of the Ancients, which
had appeared in Great Britain, France, and Germany at a period more recent.

Amongst the latter may be particularised the two
Paris editions of Pliny, the one in 10 vols., published in 1827, under the auspices of Cuvier, Des-
fontaines, and other eminent French naturalists,
the other, with a French translation, in 20 vols.,
undertaken by Pancoucke in 1829—37, and edited
by Grandsagne, with the assistance of several well-
known naturalists, but of which the botanical part
was chiefly contributed by M. A. L. à Fee; and
also the English translation of the same Author by
Dr. Bostock and H. T. Riley, Esq., brought out by
Mr. Henry Bohn in 1855, which is enriched with
a number of useful explanatory notes, for which
the Editor has been greatly indebted to the French
author last mentioned.

I may also mention Sprengel's *Historia Rei Her-*

bariæ ; the *Flore de Théocrite* and *Flore de Virgile* of
the same Mons. Fee; the *Flore Poétique Ancienne*,
by J. B. Du Molin, Paris, 1856 ; Billerbeck's *Flora
Classica*, Leipsic, 1824 ; Dierbach's *Flora Apiciana*,
Heidelberg, 1830 ; Fraas, *Synopsis Plantarum Floræ
Classicæ*, München, 1845 ; and Lenz, *Botanik der
alten Griechen und Römer*, Gotha, 1859.

With the assistance of these and other appliances,
it has been my endeavour to affix modern synonymes
to as many as possible of the Trees and Shrubs
alluded to by Classical writers, although in consequence of the vagueness of their descriptions, and
the loose manner in which they noted the characters of the plants that came under their observation, I have found it scarcely within my power,
except in the case of a few conspicuous and im-
portant species, to do more, than point out with
some degree of probability the Natural Family, or
at most the Genus, to which the classical desig-
nation appeared intended to apply.

Upon the whole, I venture to hope, that this little
Work, in conjunction with the " Lectures on Roman
Husbandry " which preceded it, will be found to
embrace an identification of a greater number of
Greek and Roman plants than is contained in any
former English Publication ; although I must in
fairness inform my readers, that a still larger list
is included in more than one of those German
writings which have been already referred to in

this Preface, and that with regard to Pliny, the
French edition in 20 vols., as well as the English
Translation referred to, contain a reference to many
which, being herbaceous, have not fallen within the
scope of the present volume.

CONTENTS.

LECTURE I.

LECTURE II.

LECTURE III.

LECTURE I.

LECTURE II.

LECTURE III.

LECTURE IV.

ON THE TREES AND SHRUBS OF THE ANCIENTS.

LECTURE I.

ON TREES PRODUCING FRUIT—ACORNS—AND RESIN.

HAVING in my former Lectures on Rural Economy included an account of the gardens of the ancients and of the plants therein cultivated, I propose on the present occasion to introduce to your notice the trees and shrubs which constituted their forests, or were admitted into their pleasure-grounds and plantations.

For this Pliny is our principal authority, as in his 12th and subsequent books he enters at large into the subject of trees and forests, regarding them as the most valuable presents conferred by Nature upon mankind. It was from the forest, he says, that man drew his first aliment, by the leaves of the trees his cave was rendered more habitable, and by their bark his clothing was supplied; indeed, even in his day, nations existed which had not advanced beyond this primitive condition.

Now it is worthy of remark, how many of the trees that we know to be well adapted for the climate of Italy, had nevertheless, according to Pliny's account, been traced to a foreign source.

Amongst Fruit-trees, we are assured that the Cherry, the Peach, the Quince, the Damson, the

B

Jujube, the Pomegranate, the Apricot, as well as the Olive, and probably even the Vine, were of foreign origin.

Thus the Cherry was brought from Pontus by Lucullus after his victory over Mithridates; the Peach, or *Malus Persica*, was introduced from Persia; the Quince, *Malus Cydonia*, from Crete; the Damson, *Prunus Damascena*, from Damascus; the Olive, in the reign of Tarquinius Priscus, from Greece; and even the Vine, though considered by Pliny as indigenous to Italy, probably found its way there from the same country.

As to the Plum, the wild variety, *Prunus insititia*, is indigenous in Italy, but the Apricot was imported from Armenia, as was the Jujube tree from Syria.

The Pistachia tree, *Pistacia vera, L.*, which yields the nuts so common in the south of France and in Italy, was introduced from Syria by Vitellius, as late as the reign of Tiberius, and from the same country came the Carob tree, *Ceratonia siliqua*, now abundant everywhere.

From Syria was derived the common Fig, *Ficus carica;* from Egypt the Sycamore Fig, *Ficus sycamorus;* and from Carthage the Pomegranate. Even the Medlar did not exist in Italy at the time when Cato wrote.

The only native fruit-trees therefore were, perhaps the Mulberry, *Morus* of Pliny, which abounded in Italy, though not known in early Greece, and therefore is by some believed to have been imported from the East; and more certainly the Apple, the Pear, the Plum, and the Sorbus; of which latter Pliny enumerates four varieties, one of which with fruits round like the apple, may be the *Sorbus domestica* of modern botanists—a tree which, though not common, is considered as indigenous even in the northern parts of Europe, and yields a fruit somewhat resembling the Medlar.

As to the other fruits, they appear, like domestic animals, to have followed man in all his migrations; and it becomes a problem, for those who maintain the late introduction of our species upon the earth, to account for so many useful vegetable productions being known already at the period of the earliest historical records, considering that now, when the whole earth has been ransacked by modern travellers, only the Maize, the Potato, and the Tobacco amongst herbaceous plants, and the Tea and Coffee amongst shrubby ones, have been added to the list of articles of daily consumption since the date of the discovery of America.

Yet when we consider, that out of the 200,000 species of plants supposed to exist upon the earth's surface, not more than a few hundreds are calculated to afford to the human race useful products, (except indeed as timber-trees or for fuel, which could not be easily transported from one country to another,) it is contrary to the doctrine of chances, that many of those trees which minister to man's wants or luxuries, should have been found growing in any one of the countries in which he was induced to settle. Even of those genera which afford edible fruits, not more than one or two species possess the qualities which fit them for man's subsistence; and in the natural course of events, centuries must have elapsed before their useful qualities were discriminated, and the species selected had multiplied sufficiently to be diffused over the countries even in which they were indigenous.

The trees above mentioned, indeed, although exotics, are perfectly well adapted, not only to Italy, but even to countries much less favourably circumstanced as to climate, but the same cannot be said of the *Malus Assyriæ* or *Malus medica*, to which Pliny alludes.

This tree has been supposed by Sprengel and others to be the Citron, or *Citrus medica*, but a late writer, Fraas, contends that it agrees with the Shaddock, both because this tree is a native of the East, whereas the Citron comes from North Africa; and also because it is used, as the *Malus Assyriæ* was, for perfuming clothes by the aromatic fragrance of the oil with which it abounds. Be that as it may, the fruit was not eaten, being, as Virgil describes it, " *Tristis succi, tardique saporis.*"

The true Orange, on the contrary, does not appear to have been introduced into Italy till the ninth century after Christ. Alphonse Decandolle[a] regards it as a variety of the Citron, which according to him is a native of China, and was gradually transported to the West by the Arabians and others. The

[a] *Geographie des Plantes.*

Crusades probably assisted in its dissemination through Europe.

The Golden Apples of the Hesperides, which some have supposed to be Oranges, seem rather to have been some variety of the Apple tribe.

No well-authenticated account is given by the prose writers of antiquity of the Citron being used as a fruit, and when Virgil introduces the shepherd Menalcas sending " *aurea mala*" to Amyntas, the poet could not have intended Oranges, which certainly were not cultivated by rustics at the time he wrote :—

> " Quod potui, puero sylvestri ex arbore lecta
> Aurea mala decem misi, mox altera mittam [b]."

In alluding to the various kinds of Indian Pepper, Pliny mentions one which is grown in Italy, larger than a Myrtle, and not very unlike it in appearance. Sprengel imagines it to be the *Daphne Thymelæa*, but this is uncertain.

The Date Palm is also spoken of by Pliny as having been introduced in his time into Europe, and it is interesting to find that those on the coast of Spain at that period bore fruit, as at present the only place in Europe where ripe Dates are grown is Elche, in Valencia. In Italy they are stated to have been barren, which agrees with our present experience, for in the most favoured climates in Italy, such as the Riviera, north of Genoa, though the Date Palm grows remarkably well, and rises to a great height, it yields no dates. This is the case even with those at Bordighera, on the road to Nice,

[b] Second Eclogue.

where plantations of Palm-trees extend almost
to the dimensions of a forest, being cultivated to
supply the churches of Italy with branches for
Palm Sunday.

The Walnut is the largest of the class of trees
mentioned by Pliny as producing edible nuts. Only
one species, the *Juglans regia*, appears to have been
known in ancient days, and from the names given
to it of *Persicum* and *Basilicum*, Pliny infers that it
originally came from Persia. Even the Filbert,
which from its abundance in the district round
Abellinum (now Avellano), in Campania, was called
Nux Abellina, came originally from Pontus, and was
thus sometimes called *Nux Pontica*.

The Almond, *Amygdalus communis*, now so common in Italy, seems hardly to have been known in
the time of Cato, unless he alludes to it by the
name of the Greek nuts, which from the epithet
acriores applied to them, may seem to have been
intended for the Bitter Almond. Columella speaks
of the Almond by the name of *Nux Græca*.

I next proceed to describe the Forest trees noticed
by ancient writers. Pliny divides them into two
classes, the glandiferous and the pitch-bearing; the
former including all the Catkin-bearing trees, or
Amentaceæ, known to the ancients, the latter most
of the *Coniferæ*.

We shall see, however, that there are many which
cannot be included in either division.

The glandiferous seem to be comprehended under
the common name of *Quercus* or *Robur*. Thus Pliny

speaks of the great Hercynian Forest as consisting of *Robora* contemporary with the creation of the world, and of gigantic size, respecting which he reports many marvels[c].

The Beech, from its possessing a kind of acorn, is ranked amongst glandiferous trees, and therefore amongst Oaks. It appears to have been indigenous in the mountainous parts of ancient Europe, and to have spread gradually towards the West, for it was not known in Holland, nor probably in England or Ireland, at the time of the Norman Conquest.

It was not the $\phi\eta\gamma\acute{o}s$ of Theophrastus, who speaks of our Beech[d] under the name of 'O$\xi\upsilon\eta$, but it was the *Fagus* of Pliny, whose description, both of its leaves and its fruit, agrees sufficiently well with our Beech, and entirely differs from that given of the Oak genus.

The Chesnut, for the same reason and with greater propriety, is classed amongst the Acorn tribe, but from the value of its nut as an article of food, it is placed by Pliny amongst Fruit-trees. It was applied to the same uses then as now, being not only roasted entire, but also ground into a sort of *polenta* or flour, for the nourishment of the peasantry. The best were those grown about Naples and Tarentum.

[c] Cæsar, *Bell. Gall.*, lib. vi. c. 26, mentions the Reindeer, *Bos cervi figura;* the *Alces*, the Elk ; and the *Urus*, the Wild Bull, *Bos priscus ;* all three animals extinct in that country, as living in the Hercynian Forest. There is a figure of each of them in the Delphin Edition of Cæsar.   [d] Lib. iii. 10.

Pliny states that the tree was introduced from Sardis in Pontus, and hence was called the Sardian Acorn; but its general distribution over all the warmer portions of Europe might rather lead us to regard it as indigenous. It forms indeed extensive forests, extending along the south of France, through Italy and Dalmatia, and re-appears in Asia Minor, where it supplied Xenophon's whole army with food in their retreat along the borders of the Euxine.

It has been supposed, that it was at one time so abundant in England as to be regarded as native, and the large beams of our old cathedrals and mansions have been said to be constructed of it; but no plant can be considered as indigenous to a country which does not ripen its fruit in it, and this can scarcely be said to be the case generally with the chesnut in England.

Those, however, who do not think this argument conclusive, may see the subject discussed in the 59th volume of the Phil. Trans. by Daines Barrington, and in the 61st by Dr. Ducarel, the former of whom takes the negative, the latter the affirmative side of the question.

Pliny enumerates twelve species of Oak as known to him, but amongst these is the Beech, which we place in a distinct genus. The remaining eleven are as follows:—

1. Quercus.	7. Suber.
2. Robur.	8. Hemeris.
3. Esculus.	9. Ægilops.
4. Cerris.	10. Latifolia.
5, 6. Ilex—two kinds.	11. Haliphlœos.

Now the following species of European Oaks are distinguished by modern Botanists, arranged after the plan of Decandolle in his late revision of the family of Cupuliferæ [c].

Genus Quercus.

Divided according as they ripen their seeds in the first year, or in the second; and then again according as their leaves are deciduous or persistent.

Oaks ripening their seeds the first year, and with deciduous leaves.

	Locality.
Quercus robur, *British Oak*, including as varieties— Sessiliflora, Pedunculata, Pubescens.	Found throughout Europe.
———— Esculus, *Italian Oak*.	Chiefly in Tuscany.

Ditto with leaves persistent.

Q. ilex, *Holm Oak*, including *Gramuntia* and other varieties.	South of Europe.
Q. suber, *Cork-tree*.	South of Europe.

2nd. Oaks ripening their seeds the second year, and with deciduous leaves.

Q. cerris, *Turkey Oak*, including as a variety *Q. ballota*, found also in Syria [f].	Greece and Turkey.

[c] See Gay, *Ann. des Sc. Nat.*, 4th Series, vol. vi., and Endlicher, *Genera*, Suppl.   [f] Hooker.

Ditto with persistent leaves.

Q. pseudo-suber, *false Cork-tree.*	Italy.
Q. Ægilops, *Valonia* [g].	Levant.
Q. coccifera [h], *Kermes Oak.*	South of France.

The only Oak mentioned by Hooker as occurring in Syria, which is not indigenous in Europe, is *Q. infectoria*, abounding in red galls, and placed on that account by Spach, in a distinct section, Gallifera, under which head Endlicher includes also *Q. humilis, Q. alpestris,* and *Q. hispanica* or *lusitanica,* which latter, however, Gay places under *Cerris.*

Let us now endeavour to identify these species with the varieties of Oak enumerated by Pliny.

The first two, *Quercus* and *Robur,* are often given by ancient writers as generic terms for the Oaks, or acorn-bearing trees ; but in his 16th Book, in which he mentions the several varieties of the genus *Quercus,* Pliny, as M. Fee thinks, would apply the term *Quercus* and *Robur* to the two varieties of *Quercus robur,* the *sessiliflora* and *pedunculata,* recognised by modern botanists.

As, however, the distinction between these two varieties is one of a botanical kind, hardly likely to be recognised by a Roman naturalist, being dependent on the fruit being in the one case attached

[k] *Ægilops* is found in Syria, and its acorns, which are often very large, are eaten by man. Hooker, Linn. Tr., v. xxiii.

[h] In Syria the most common Oak is *pseudo-coccifera,* which Hooker thinks may be a variety of *coccifera,* Linn. Tr., vol. xxiii. Abraham's Oak at Mamre, of which Hooker gives a drawing, is of this species.

directly to the branch, and in the other seated on
a footstalk, I must suspend my opinion on this
point until further proofs are forthcoming, and
for the present shall be content with supposing,
that both these names apply to species or varieties
of Oak of the largest size, and of the widest distribution throughout Europe.

We next come to his *Esculus*, a name applied
by the moderns to a species of Oak found especially
in Tuscany, and furnishing acorns so sweet as to
be much eaten by the peasantry. In olden time
the acorn, as we read in Homer and Hesiod, was
the common food of the Arcadians, so that the
tree which yielded it must have been different
from the Oak of this country, the fruit of which
is bitter and distasteful. Yet Pliny tells us, that
the best and largest acorn is that which grows
upon the *Quercus*, and only the next to it in
quality on the *Esculus*.

Some writers identify the latter with the Beech,
regarding it as a synonym for the *Fagus* of the
Latins, and the φηγός of the Greeks, which latter
name they suppose was derived from the fitness
of its fruit for food to man. The Beech-nut, indeed, is sweet and not unpalatable, so that in the
south of Europe it is eaten by man as well as
by beast.

Virgil, in his 2nd Georg., describes the *Esculus*
thus :—

> " Nemorumque Jovi quæ maxima frondet
> Esculus."

Distinguishing it from *Quercus* by adding—

"Atque habitæ Græcis oracula quercus."

He also describes it as deeply rooted—

"Quantum vertice ad auras
Æthereas, tantum radice ad Tartara tendit."

A description, however, which he transfers to the *Quercus* in the 4th Æneid.

But Virgil and Horace do not seem to speak of the same tree as Pliny, for his *Q. esculus* is a small tree, whilst their descriptions apply to one of gigantic proportions.

Martin[i] justly remarks, that although the tree may be the φηγός of Theophrastus, it cannot be the *Fagus* of Pliny, for the latter describes the *Fagus* as having its gland included in a triangular covering, and with a leaf small and very light, resembling the Poplar ; remarking that the different species of Oak have a gland, *properly* so called.   Ovid also and Virgil both describe the *Esculus* as having *large* leaves.   It seems to have been less common in Italy than the *Quercus*, although Horace speaks of extensive forests of it in Apulia.   I cannot, therefore, submit to Martin's view, that it is the sessile variety of the common Oak, but regard it as the Tuscan Oak, *Quercus esculus* of the moderns.

Pliny and Theophrastus both describe an Oak under the specific name of *Hemeris*, ἡμερίς.   The latter term was evidently applied to it by the

---

[i] Notes on Georg. 2.

Greeks on account of the mild or sweet character of the acorn, as contradistinguished from the
bitterness of other varieties, just as we speak of
the sweet Chesnut to distinguish the fruit of the
Spanish from that of the Horse Chesnut.

Pliny states, that although a small tree, it bears
the largest acorn of any of the Oak tribe, and
also yields the best gall-nuts.

Sprengel seems to regard it as the *Quercus cerris*,
but if so, it must be the tree which Decandolle[j]
distinguishes as *Q. toza*, and which, as he suggests,
may be a dwarf variety of *Cerris*, a tree too lofty
in its ordinary character to agree with the *Hemeris*
of Pliny.

The *Ilex* of that author and the Πρίνος of
Theophrastus, are evidently our *Quercus ilex*, or
Holm Oak.

Pliny enumerates two varieties of this evergreen
tree, but the one with a leaf not unlike the Olive,
called *smilax* by the Greeks, and *aquifolia* by the
Latins, is not an Oak at all, but probably our common Holly, the *Ilex aquifolium* of modern botanists,
belonging to the family of *Rhamneæ*. This plant,
however, only bears a berry, and has no fruit resembling the Oak, a circumstance which may throw
a doubt upon its identification, and lead us to believe that Pliny must have intended some particular
kind of evergreen Oak.

Under the term *Ilex*, Pliny, however, confounds
with the Holm that species of evergreen Oak

[j] Fl. Fr. Suppl.

indigenous to the south of Europe, which from
its being subject to the attacks of the insect called
the *Coccus ilicis*, occasioning by its puncture a
little red gall, known as Kermes, is denominated
by modern botanists *Quercus coccifera*.

This scarlet berry, or *grain*, as Pliny calls it, was
highly prized in dyeing until the discovery of the
cochineal insect, which from its brighter hue has
superseded it, and in the time of the Romans it
was a great article of commerce, furnishing to the
poor of Spain the means of paying one-half of the
tribute exacted of them.

Pliny, however, has committed the double inac-
curacy, of confounding the tree which bears it with
the *Ilex*, which indeed is not surprising, as they are
both evergreen species of the same genus, and also
with the *Aquifolium*, which is, as already remarked,
our common Holly.

The *Haliphlœos* is stated as being the worst of
any, both for firewood and for timber, having a remarkable thick bark, and a trunk, which although
of considerable size, is for the most part hollow
and spongy, so that it rots even whilst the tree is
alive. It rarely bears acorns, and what there are,
have a taste so bitter and forbidding that no animal
will touch them.

The description seems to correspond best with
the *Q. pseudo-suber* of the present day, a native of
the mountains of Tuscany, Spain, and Barbary, the
bark of which is corky, although in a less degree
than the true Cork-tree.

Its name is taken from ᾿Αλίφλοιος, a kind of Oak described by Theophrastus much in the same terms as those which Pliny uses, and deriving its name from ἅλς, 'salt,' and φλοιός, 'bark.'

The *Q. cerris* of Pliny would seem to be the Turkey Oak of modern writers, although there is much confusion in the account given of it, and we have seen that some imagine the *Hemeris* to be the *Cerris* of botanists, which however does not accord with the bitter character of its acorn.

The *Suber* of Pliny is the φελλός· of Theophrastus, both of which are described in such a manner by the Greek as well as by the Roman naturalist, as to identify them with our Cork-tree.

The only circumstance which might lead us to doubt this, is that in one passage Theophrastus states that it sheds its leaves annually, whereas our Cork-tree is an evergreen. In another passage, however, he calls it, as Pliny and other Roman writers have done, an evergreen. Commentators have tried to get over this difficulty, by pointing out that there is actually a variety of the Cork-tree, which sheds its leaves in April, and which has been observed near Bayonne [k]. The time of shedding the leaves is, indeed, not characteristic of a species, for the Lucombe Oak [l], which passes for an evergreen, as its leaves remain on all the winter, is

---

[k] *Clusius, Plant. rar. Hist.,* lib. i. c. 14, *and Miller's "Gardener's Dictionary."*

[l] Pliny, lib. xvi. c. 21, records as a marvel, that a single Oak (*Quercus*) existed in the territory of the Thurii, where Sybaris formerly stood, which never lost its leaves, and did not bud till Midsummer.

a variety of the Turkey Oak, the leaves of which are deciduous.

Cork was employed by the ancients for many of the same purposes for which it is used at present, as for the soles of sandals for females, both to keep out wet and to make them appear taller. Also for swimming-jackets[m] and for floats.

But though it was sometimes employed for stopping the holes of casks, its application to this object does not appear to have been common, as pitch, clay, gypsum, and even oil or honey, were used for excluding air from their liquors. Indeed until glass bottles were in use, corks would be rarely wanted for the purposes to which they are mainly applied at the present time.

The *Ægilops* is stated to be the loftiest tree of all, and to be attached to wild, uncultivated spots, producing, according to Theophrastus, the hardest and best timber of any.

It is probably the species now known as *Ægilops*, the finest and tallest of the Oaks that occur in Greece, and the one which affords the acorns used in dyeing, called Valonia.

The *Latifolia* is only described by Pliny as being next in height to the *Ægilops*, so that it seems difficult to identify it. It is called πλατύφυλλος by Theophrastus: query *Quercus Tournefortii Pers.*, which is generally regarded as a synonym of *Q. cerris*, or the Turkey Oak.

---

[m] *Plut. in vitâ Camilli.*

Pliny enumerates the various uses to which trees of the glandiferous or cupuliferous family were applied in his time.   .

We have already alluded to the Kermes employed as a dye, and have also noticed the galls, found, as Pliny says, on all species of Oak, and probably indeed on most. Of this the *Hemeris* and the *Latifolia* produced the best kinds. That the galls were occasioned by the eggs of an insect deposited upon the leaf or bark of the tree, the ancients do not appear to have been aware, although Pliny notices, that a kind of gnat is produced in the veins of the leaves, and comes to maturity just in the same way as the ordinary gall-nut does.

Pliny also notices the edible mushrooms which grew at the roots of the *Quercus*, and which were highly esteemed in cookery; and also the *Agaric* of the Oak, a fungus which is used as a styptic, and furnishes the description of tinder called in modern times Amadou.

He also mentions *Cachrys*, as a useful medicinal product obtained from the *Robur*, the nature of which is not known to us, though probably either a fungus or some kind of excrescence.

The acorn of the Beech, when given to swine, makes them brisk and lively, and renders their flesh tender, and easy of digestion. Next to this stands the acorn of the *Cerris*, but that of the *Ilex* is less wholesome to them.

The acorn of the *Quercus* and of the *Latifolia* was in much esteem as an article of food for man.

The Oak tribe, too, stood highest in the list of timber trees, and the best shingles for covering the roofs of houses were obtained from the *Quercus robur*.

Its bark was also applied to a variety of uses, and especially to covering the roofs of houses; that of the Cork-tree, as we have seen, to most of the purposes for which in modern times it is made available.

I will conclude this account of the Oak tribe by classifying the Oaks known to the ancients according to the system proposed by the younger Decandolle :—

QUERCUS.

Fruits.		Leaves.	Ancient names.	Modern names.
Appearing the 1st year.	{	Deciduous	Robur	Robur
		,, ,,	Esculus	Esculus
		Persistent	Ilex	Ilex and Coccifera
		,, ,,	Suber	Suber
The 2nd year	{	Deciduous	Hemeris	Toza
		,, ,,	Cerris	Cerris
		Persistent	Halephlæus	Pseudo-suber (?)
		,, ,,	Ægilops	Ægilops

I shall now proceed to notice another class of trees, distinguished by Pliny under the name of pitch-bearing, comprehending those now known as the *Coniferæ* or Fir tribe.

These Pliny divides into *Abies* and *Pinus:* and modern botanists, having separated the *Abietinæ* into two groups — namely, the one with leaves solitary or in two ranks, the other in clusters of two, three, or five each—place the former under the head of *Abies*, and the latter under that of *Pinus*.

But we must not suppose that Pliny contemplated any such division. On the contrary, the Spruce Fir, which stands as the very type of the genus *Abies*, is not indigenous either in Italy or Greece. Loudon, therefore, and other botanical writers, are in error, when they regard the *Abies* of the Latins as the Spruce Fir of Northern nations.

In order to ascertain what kind of tree Pliny meant by the term *Abies*, and Theophrastus by the corresponding one Ἐλάτη, our best method will be to inquire, in the first instance, what are the species indigenous in Greece and Italy.

In Greece Sibthorp enumerated the following :—

1. *Pinus sylvestris*, Scotch Fir, which he states to be found in the mountains of Bithynia. As this, however, has not been confirmed by succeeding travellers, it seems doubtful whether he may not have confounded with it the Corsican Pine, *P. Laricio*, which though omitted by him, is recognised by other botanists (Lambert, *Genus Pinus*, Gussone, *Flora Sicula*), as existing in all the southern parts of Europe.

2. *Pinus pinea*, Stone Pine, Πίτυς of Dioscorides [n], met with on the sandy shores of Western Peloponnesus.

3. *Pinus maritima*, Maritime Pine, Πεύκη of Dios-

[n] i. 86.

corides, found everywhere in the sandy flats of Greece, and especially in Elis. It is probably the same as *P. halepensis*, which Sibthorp omits, but which is stated by other writers as the commonest Fir in Greece, from the sea-shore to a height of about 3,000 feet above it.

4. *P. picea*, or *Abies pectinata*, the Silver Fir of modern botanists, and the Ἐλάτη of Theophrastus, which is met with commonly on the loftier mountains of Greece.

In Italy the same species occur, and in addition to them the *P. pinaster*, or Cluster Pine, is abundant as far south as Genoa, where it gives place to the *Pinus halepensis* already noticed, and, according to Tenore°, to three others, namely, *P. brutia, pumilio*, and *uncinata*.

In the Alps, too, and the south of France, the *Pinus Mugho* or *uncinata*, and *Cembra* are abundant ; so that the Roman writers may have had in their eye five more species of Fir than those occurring in Greece.

Now in order to prove which of the species above assigned is the one designated by Pliny under the name of *Abies*, and by the Greeks under that of Ἐλάτη, let us consider the properties assigned to that tree.

1. It was especially useful in ship-building. Hence in Euripides[p] Ἐλάτη is used for a ship.

2. It grows chiefly on the summits of mountains.

3. It resembles in form the *P. picea*.

° *Flora Neap*.   [p] Phœn. 208.

4. It is chiefly used for beams, and other purposes for which solidity is requisite.

5. It gives out so much resin, that the quality of the wood is often impaired by the quantity emitted, even the warmth of the sun being sufficient to cause an exudation; whereas the same process is even serviceable in the case of the *Picea*.

6. Lastly, it is inferior in the quality of its timber to the last-named species.

Now of the *Pinuses* above enumerated as existing in southern Europe, the *Abies pectinata* is the one which seems best to accord with the above description, especially when we add, that Pliny [q] describes it as having its leaves indented like the teeth of a comb, which may be regarded as expressive of one of the generic distinctions between the *Abies* and *Pinus* of modern botanists.

But we must not expect from this Author, or indeed from any of those of antiquity, the same precision as is demanded from modern Botanists in such matters. Probably the two lines in Virgil's 7th Eclogue, v. 65,—

> " Fraxinus in silvis pulcherrima, Pinus in hortis,
> Populus in fluviis, Abies in montibus altis,"—

express the amount of discrimination which the Romans exercised in such matters; so that, not only the *Abies pectinata*, but any other resinous tree, with narrow pointed leaves, growing in mountainous

----

q Lib. xvi. c. 38.

places, attaining to a great height, and serviceable for timber, would have been included by them under the name of *Abies*.

Thus when Cæsar[r], in describing the productions of Britain, says, "Materies cujusque generis, ut in Galliâ est, *præter fagum atque abietem*," he must have alluded to the Scotch Fir, the only species of the tribe indigenous in this country.

Palladius[s], however, may refer to the Spruce Fir under the name of the *Abies gallica;* the timber of which, he says, is smooth and hard, and is of a character especially durable for buildings, except under water, provided only that it be not exposed to wet just after it has been felled.

Now the Spruce, though indeed not found in Italy, is indigenous in Dauphiny and throughout the Pyrenees (*Flora Gallica*), and its resemblance to the Silver Fir might easily lead to its being confounded with it, although the specific name of *Gallica* would seem to shew that the ancients recognised the distinction between the two.

A difficulty, indeed, arises from the distinction set up by Theophrastus between the male and female 'Ελάτη.

From this recognition of the distinction of sexes in these and other instances, it has been inferred, that the ancients possessed some glimpse of the modern doctrine on the subject which has been established so fully by the researches of botanists of this and the preceding century.

<hr/>

[r] *Bell. Gal.*, lib. v. c. 12.   [s] *Novenbris* xv. 11.

It cannot, indeed, be denied, that in certain diœcous plants,—those, I mean, in which the sexes are on different *trees*,—the ancients were aware of this great truth.

Aristotle[t] states, that if we shake the dust of the male palm over the female, her fruits will quickly ripen ; and moreover, that when the wind sheds this dust upon the female, her fruits become developed just in the same manner as if a branch of the male had been suspended over her.

Theophrastus also, treating on the same subject, says, " They bring the male to the female palm, in order to make her produce fruits, proceeding in the following manner. When the male is in flower, they select a branch abounding in that downy dust which resides in the flower, and shake it over the fruit of the female. This operation prevents the fruit from becoming abortive, and brings it soon to perfect maturity."

After quoting these two passages from the two most accurate naturalists of ancient days, I shall not insist upon the more vague statements of Herodotus,—who, in alluding to the Date-palm, informs us, that the necessity of bringing the male into the proximity of the female blossoms was known to the Babylonians,—seeing that this writer compares the practice to what is called the caprification of the Fig.

Now the latter operation consisted in bringing from the forest the branches of the wild fig-trees

[t] *De Plantis.*

and placing them over the cultivated ones, in order to cause them to ripen their fruit sooner, an effect in this instance merely ·due to the multitude of insects of the *cynips* tribe, which were thus brought into contact with the fruit of the cultivated variety, and which pierce with their little proboscides the membrane which incloses the seeds, thus by their stimulus causing them to ripen sooner, so as to allow time for a second crop.

Accordingly, in this case the same effect may be brought about by simply puncturing the figs with a sharp instrument, and introducing a drop of oil into the puncture to keep it open.

This process, therefore, bears no real analogy to the influence of the pollen upon the stigma of a flower; so that it would appear that Herodotus had a very vague notion of the nature of this latter operation.

Neither does Pliny appear to have advanced beyond the knowledge possessed by the writers already quoted; for although he states, that naturalists admit a distinction of sex, not only in trees, but in herbs, and *in all plants* whatsoever, yet in proof of this position he only quotes the case of the Date-palm, respecting which he offers a description of the same nature as that given us by Theophrastus, although one couched in more poetical phraseology.

Thus he says ", "that in a forest of natural growth the female trees will remain barren if

" Lib. xiii. 7.

isolated from the male, and that several females may be seen bending towards the latter, with a foliage of a softer character. The male tree on the contrary, bristling with erect leaves, fecundates the others, by its presence, by its exhalations, and even by the dust it emits. And when it is cut down, the females, reduced to a state of widowhood, become barren. So well, indeed, is this sexual union recognised as taking place, that the idea has arisen, of securing the act of impregnation by man's agency, the blossoms from the male trees being gathered for that purpose, and even sometimes nothing more being done, than to sprinkle the dust taken from the same over the female trees [x]."

---

[x] This idea has been elegantly worked out by a Dutch poet of the seventeenth century, Fontanus, in some lines which may perhaps have suggested to Darwin his "Loves of the Plants:"—

"Brundusii latis longe viret aurea terris
    Arbor, Idumæis usque petita locis
Altera Hydruntinis [y] in saltibus æmula palmæ
    Illa virum referens, hæc muliebre decus.
Non uno crevere solo, distantibus agris
    Nulla loci facies nec socialis amor
Permansit sine prole diu, sine fructibus arbor
    Utraque, frondosis et sine fruge comis
Ast postquam patulos fuderunt brachea ramos
    Cœpere et cœlo liberiore frui,
Frondosique apices se conspexere, virique
    Illa sui vultas, conjugis ille suæ
Haurire et blandum vinis sitientibus ignem
    Optatos fœtus sponte tulere suâ.
Ornarunt ramos gemmis, mirabile dictu !
    Implevere suos melle liquente favos."

[y] Now Otranto.

This is the only passage, so far as I am aware, in which Pliny adduces any facts in favour of the existence of sexes in plants, and they are taken exclusively from the case of diœcous ones.

As to the well-known lines of Claudian, so often quoted in support of the idea that the ancients possessed a knowledge of the modern system of the sexuality of plants, where the poet says,—

> " Vivunt in venerem frondes, omnisque vicissim
> Felix arbor amat, nutant in mutua Palmæ
> Fœdera, populeo suspirat Populus acta
> Et Platani Platanis, Alnoque adsibilat Alnus,"—

all we can say is, that they are a poetical amplification only of the idea thrown out by the prose writers alluded to, and afford no additional evidence in favour of its truth.

In coming to the conclusion, therefore, that *all* plants are similarly circumstanced, Pliny seems to be guided by analogy merely, nor is there reason to infer that he possessed any knowledge of the distinct function performed in the case of flowering plants in general by the stamens and pistils where the two are conjoined.

At any rate, the trees to which the ancients applied the terms *male* and *female* are not such as shew this distinction of sexes on separate individuals. Thus no plant to which the term ἐλάτη, or *Abies*, can by possibility be applied, possesses male flowers on one, and female ones on another separate and distinct individual. Accordingly, we are driven in these cases to explain the epithet applied in some other manner, quite apart from any consideration of sex.

Liddell and Scott accordingly consider the so-called male plant as the Spruce, and the female one as the Silver Fir.

This identification, however, would seem irreconcileable with the fact, that the former tree is not indigenous either in Greece or in the Levant.

Lenz, on the contrary, lays it down that the male Fir is the *red*, and the female the *white* variety of the Spruce; to which idea of course the same objection applies.

All we can say for certain on this subject is, that the one called the male Fir was the loftier, hardier, and handsomer tree of the two. It is, therefore, probably the one which Homer alludes to as soaring to the sky, (οὐρανομήκης,) and describes as growing in the Isle of Calypso, along with the Alder, the Poplar, and other lofty trees :—

> " ἦρχε δ' ὁδοῖο
> Νῆσου ἐπ' ἐσχατιῆς, ὅθι δένδρεα μακρὰ πεφύκει,
> Κλήθρη, τ' αἴγειρός τ', ἐλάτη τ' ἦν οὐρανομήκης ᶻ."

> " On the lone island's utmost verge there stood
> Of poplars, pines, and firs, a lofty wood,
> Whose leafless summits to the skies aspire,
> Scorched by the sun, or seared by heavenly fire."

The word ἐλάτη indeed is used by Dioscorides in quite a different sense, namely, to indicate the young

ᶻ Odyss. v. 239.

fruit of the Palm, in which sense Galen also employs it, applying however the term adjectively [a] for the resin which flows from the Fir, and substantively [b] for a plant of quite a different description from the ἐλάτη, probably the *Antirrhinum orontium* of Linnæus.

Theophrastus points out a difference between this and other trees, in respect to the circumstance that it is destroyed, by having its top cut off, but not by the removal of the entire tree, trunk, branches, and all, so as only to leave the root and the stump at bottom standing.

In the former case, he says, the tree speedily dies, whilst in the latter it will put forth again fresh shoots.

With regard to the degree of danger incurred in lopping Firs modern writers are divided, all however agreeing to regard it as a prejudicial practice [c].

Of the tendency of the tree in some rare cases to put forth fresh branches when cut nearly down to the ground, an instance is given with respect to a forest situated in the Pyrenees, in the work of a Mons. Leroy, Paris, 1777.

It seems that in this case the forest was wholly made up of shoots that had pushed forth from the stumps, after the trees had been felled nearly to the ground, no less than twelve branches being sometimes known to have sprung from the same trunk.

[a] Lib. i. c. 92.   [b] Lib. iv. c. 2.
[c] See Selby on Forest Trees.

Theophrastus also remarks, as a peculiarity of this tree, that when a branch is broken or cut off, a protuberance forms round the cicatrix, which is black, and of such hardness, that the Arcadians made cups out of it; adding, that the softer the wood is, the harder is the protuberance.

Modern writers speak of this as a swelling in the bark of the tree, attributing it to an insect, which by irritating the vegetable tissue determines a flow of resinous matters to the part[d].

The next species of Fir noticed by Pliny is the *Pinus*, the πίτυς of Dioscorides. The former writer states, that it was common about Rome in his time, that its nuts or seeds were eaten, and that it sends out branches from the top. These particulars would lead us to identify the tree with the *Pinus pinea*, or Stone Pine, which forms so conspicuous a figure in the Roman landscape; and this suspicion is confirmed by a letter of the younger Pliny, who in his account of the great eruption of Vesuvius, which destroyed Herculaneum and Pompeii, compares the smoke which rose from the volcano to this Pine in point of shape. Now the vapour which ascends from a volcano resembles the Stone Pine in this respect, that it runs first in one narrow vertical trunk, and then spreads itself out laterally on all sides in a conical form. Such is the case with the *Pinus pinea*, which has no branches on the lower part of its trunk, but sends out numerous ones in an umbrella form from its summit.

[d] See Selby on Forest Trees.

The next kind of Fir enumerated by Pliny is the *Pinaster*, but whether this be the same as the species now known by that name seems doubtful.

It is called by Theophrastus πεύκη ἄγρια, and Pliny says that it is nothing else than a wild variety of Pine. He informs us, however, that it rises to a surprising height, that it throws out branches from the middle of its trunk, and that it yields a more copious supply of resin than the *Pine*.

"Many people," he says, "think that this is the same tree which grows along the shores of Italy, and is known by the name of *Tibulus;* but the latter is slender, and more compact than the *Pine;* it is likewise free from knots, and hence used in the construction of light galleys. It is also as deficient in resinous matter as the *Pinus*.

Now it seems doubtful from this description, whether the tree called *Pinaster* by Pliny ought to be identified, as has been generally done, with that so called in modern times.

It does not appear that Pliny's *Pinaster* grows near the sea, although the *Tibulus*, which has been mistaken for it, does; and the Fir so common on the sandy flats of Greece is not the *Pinaster* or Cluster Pine which has proved so invaluable in reclaiming large tracts of country on the western coasts of France, but the *P. maritima*, which is probably the same species as *P. halepensis*, the Aleppo Pine of modern authors.

We next find noticed the *Picea*, so called from the abundance of resinous juice it exudes. It loves mountain heights, and cold localities. It spreads out its branches from the very root of the tree, adhering to its sides like so many arms. Its wood is much valued for the construction of rafters, and of other articles requiring strength and durability.

This would apply to the Scotch Fir of our island, but as the latter does not extend much into the southern parts of Europe, the Corsican Pine, *Pinus laricio*, is more probably the one intended.

It seems to be the same which Theophrastus[e] calls πεύκη ἰδαῖα, and which he there distinguishes from πεύκη παραλίας, the *Pinaster*.

Next we find noticed by Pliny the *Larix*, which commentators in general consider the same as the Larch of the present day, a tree not known in Greece. It abounded in turpentine, as does our Larch, from whence the Venice turpentine is chiefly derived, but the durability ascribed to its timber is greater than the moderns are wont to attribute to the Larch.

Lastly, Pliny notices the *Tæda* as a Fir which yields more resin than any other excepting the Pitch-tree, and was on that account used for torches in religious ceremonies. The term *Pinus tæda* is applied by modern botanists to a Fir indigenous in North America, and extending from Florida to Virginia, called the Frankincense Pine, also remarkable for the quantity of turpentine which it affords.

[e] Lib. iii. c. 8.

The *Tœda* of the ancients may perhaps be identified with the *Pinus Muglo* of modern botanists,
the Torch Pine of the French, the *Pumilio*, or Dwarf
Pine of old writers, indigenous in most mountainous
parts of Europe, more especially in France and
Germany, which is also a tree highly impregnated
in turpentine.

Accordingly, we may with some degree of probability identify the species of Fir named by Pliny
and Theophrastus with the following modern
species :—

	Ancient.	Modern.
Leaves in pairs.	Pinaster.	{ Pinus halepensis, or maritima.
,,	Tæda.	—— mugho.
,,	Picea.	—— laricio.
. ,,	Pinus.	—— pinea.
in fives.	Strobus (?) [f].	—— cembra.
	Excelsa [g].	
Solitary, evergreen.	Abies.	Abies pectinata.
,, ,,	—— gallica.	—— excelsa.
Fascicled, deciduous.	Larix.	Larix.

[f] Only mentioned once by Pliny, lib. xii. 40, as used in fumigations.
It seems rather rash to identify it, as Fraas has done, with the modern
Cembra.

[g] Identified by Dr. Hooker with the *Pinus Peucè* of Griesbach,
which that botanist had noted on Mount Peristeri in Macedonia, and
had considered as intermediate between *P. cembra* and *P. strobus*.

The *P. excelsa*, so common in the Himalayas, has not been observed
nearer to Greece than Afghanistan, a distance of more than 2,200
miles.—(Jour. of the Linn. Soc., vol. viii. No. 31.)

LECTURE II.

THERE are still a few plants not placed by the
ancients under the name of *Piciferæ*, which
either belong, or bear a near relation to, those
which modern botanists regard as *conifers*. Thus
we have repeated mention of a tree called *Cedrus*,
or in Greek κέδρος, which afforded timber for
many useful purposes, and was at the same time
aromatic.

We are, therefore, naturally led to suppose that
it was a Cedar, and you will see it set down in
most dictionaries and lexicons[a] as corresponding
to that tree. Nevertheless, it is difficult to suppose
that when the writers of antiquity speak of it as
existing in Italy or Greece, they could refer to
any one of the three species of Cedar recognised
by modern botanists, namely, either to *C. Libani*,
C. Atlantica, or *C. Deodara*.

[a] See Scott and Liddell.

D

Theophrastus, indeed, mentions the large size which the κέδρος attains in Syria, in this instance alluding plainly to the Cedar of Lebanon; but neither this nor the other two known species appear to have found their way into Greece or Italy in ancient times, nor indeed are they recognised at the present day as indigenous in either country.

The Cedar of Lebanon was not introduced into England till about the time of Evelyn; nor into France till much later, namely, in 1737, when Bernard de Jussieu brought over from the Holy Land a little seedling of this plant from the forests of Mount Lebanon.

A romantic account is given of the difficulty this naturalist experienced in conveying it to France, owing to the tempestuous weather and contrary winds he experienced, which drove his vessel out of its course, and prolonged the voyage so much that the water began to fail. All on board were consequently put upon short allowance; the crew, having to work, being allowed one glass of water daily, the passengers only half that quantity. Jussieu, from his attachment to botany, was induced to abridge even this small daily allowance, by sharing it with his plant; and by this heroic act of self-sacrifice, succeeded in keeping it alive till they reached Marseilles.

Here, however, all his pains seemed likely to be thrown away, for as he had been driven, by want of a flower-pot, to plant it in his hat, he excited on landing the suspicion of the Custom-house officers, who at first insisted on emptying the strange pot, to see whether any contraband goods were therein concealed.

With much difficulty he prevailed upon them to spare his bantling, and succeeded in carrying it in triumph to Paris, where it flourished in the Jardin des Plantes, and grew until it reached one hundred years of age, and eighty feet in height. In 1837 it was cut down to make room for a railway, and now the hissing steam-engine passes over the place where it stood.

Nor is the Cedar mentioned in any of the Floras of Greece or Italy, and if it had existed in either country in the days of Theophrastus or Pliny, it could not have failed to be noticed by them for its peculiar and noble appearance.

Although called the Cedar of Lebanon, it does not seem to be so abundant in that locality as in some others. Tchihatcheff, a Russian traveller, speaks of vast forests of it on Mount Taurus in Asia Minor, whilst on Lebanon only a very few of the older trees survive; and Dr. Hooker, in 1860, counted in all no more than about 400 trees, of which only 15 exceeded fifteen feet in girth [b].

Nor do the qualities of the tree correspond to those attributed to the Cedar of the ancients. Whatever may have been the durability of the wood of which Solomon's temple was built, this

---

[b] Mr. Tristram, however, a more recent traveller in the Holy Land, appears to have discovered a new locality in the Mountains of Lebanon, where this Cedar is more abundant.

tree in other regions does not appear to be adapted
for the purposes to which in ancient times the
Cedar was applied.

Virgil[c] speaks of the *Cedrus* in common with
the Cypress as employed in house-building : —

> " Dant utile lignum
> Navigiis pinos, domibus cedrumque cupressosque."

In the 7th book of the Æneid, line 13, he alludes
to the Cedar as used for fragrant torches :—

> " Urit odoratam nocturna in lumina cedrum :"

and, in line 178, for statues, the images of the
ancestors of Latinus being carved out of an old
Cedar :—

> " Quin etiam veterum effigies ex ordine avorum
> Antiqua e cedro, ...
> Vestibulo adstabant."

But if the word *Cedrus* cannot be supposed to
have any reference to our Cedar, it is somewhat
difficult to determine what the tree intended can
have been.

That which seems best to correspond with the
descriptions given of the *Cedrus* by ancient writers
is the Juniper, and hence it may be worth while to
note down the accounts given of it by classical
authorities.

Now that which we meet with in Theophrastus
is probably the most authentic, as it is the most
detailed, of any that can be gathered from their
writings.

[c] Georg. ii. 443.

This writer remarks[d], that according to some there are two sorts of Juniper, ἀρκευθος, one of which has blossoms but no fruit, the other fruit but no blossoms, like the Fig.

And in his eleventh chapter he goes on to say that many consider that there are two sorts of Cedar, κέδρον, the Lycian and the Phœnician.

Others, as the inhabitants of Ida, know only of one kind, which is like to the ἀρκευθος; but there is a difference in the leaf, which in the κέδρον is hard-pointed and prickly, but in the ἀρκευθος softer. The ἀρκευθος, also, seems to be the loftier tree of the two.

There are people, nevertheless, who apply the name of κέδρον to the ἀρκευθος, as well as to the tree which more properly receives that appellation; or else, calling the ἀρκευθος, κέδρον, they distinguish the tree properly called κέδρον, by the name of ὀξύκεδρον.

Both have many branches and a knotty timber. The heart-wood of the κέδρον resists decay, and is red, but in the κέδρον it is fragrant, whilst in the ἀρκευθος it is not so.

The fruit of the κέδρον is yellow, of the size of that of the Myrtle, and possesses an agreeable smell.

The ἀρκευθος has a similar kind of fruit, but it is black, rough, and almost disagreeable. It hangs on the tree for a year, and then gives place to new fruit. Its bark is like that of the Cypress, but

[d] Hist. Pl. iii. 4.

rougher; both have loose and flat-growing roots. It thrives chiefly in rocky and cold situations.

Pliny[c], who, as is usual, follows Theophrastus, says that Phœnicia produces a small Cedar, very like the Juniper, of which there are two varieties, the one found in Phœnicia, the other in Lycia. But he adds, there are two kinds of large Cedar also, the one that blossoms, but bears no fruit, the other bearing no blossoms, but a fruit similar to that of the Cypress.

This latter would seem to have been the same tree to which Virgil alludes, as he goes on to state, that the wood is so durable that it has been used for making images of the gods, and that there is in a temple at Rome a statue of Apollo Sosianus in cedar, originally brought from Seleucia.

Now in his description of the two kinds of large Cedar, Pliny evidently refers to the male and female Juniper; for although it is not true that the fruit-bearing trees are destitute of blossoms, yet the latter, being small in their early stage, might have been overlooked.

But if so, it still remains a question, to which species of the Juniper tribe Pliny could have re-ferred.

We find, according to Sibthorp, five species of Juniper indigenous at the present time in Greece, and of these only three, or at most four, are iden-tified by that botanist with any of the plants named by Dioscorides.

[c] Lib. xiii. c. 11.

These are *Juniperus communis*, ἀρκεύθος μίκρα; *Juniperus oxycedrus*, κέδρος μίκρα; *Juniperus phœnicia*, βράθυς ἕτερον; *Juniperus sabina*, βράθυς.

1. *Juniperus communis* loves cold places, and is frequent on hilly places in north Italy. In Greece it is found on the highest mountains, as at Athos and Olympus. The ancient Greeks had no particular name for it, but designated it as ἀρκεύθος μίκρα.

2. *Juniperus oxycedrus*, a shrub of about six feet in height, common all along the borders of the Mediterranean. To this the *J. macrocarpa* of Sibthorp is very like, as well as *J. rufescens* of Link, which grows in the south of Europe, from Portugal to Macedonia and Thrace. The ancients called all three κέδρος or κέδρον.

3. *J. phœnicia*, or *lycia*, is a low shrub, like the Savine, of which Dioscorides considers it a variety, with a strong aromatic smell, common in Asia Minor, Arabia, and Syria.

4. *J. sabina*, indigenous on the southern and northern slopes of the Alps, and more rarely on the mountains of Greece, was termed by the ancients βράθυς. The ancients, it may be observed, called the Juniper generally by the name of Cedar; although Pliny, as we have seen, distinguishes between the two, saying that Phœnicia produces a small Cedar which bears a strong resemblance to the Juniper.

Now it is evident that none of the above correspond with the description given by Pliny of the larger kinds of *Cedrus*; those from which the beams of houses and the images of the gods were fabricated.

But the *Juniperus thurifera*, a native of Spain and Portugal, grows to the height of twenty-five feet and upwards, and the *J. excelsa*, first discovered in Siberia by Pallas, and now found to be common in Asia Minor, Syria, and the Grecian Archipelago, is still loftier [f].

Is it not probable, that one of these may have been known to the ancients ? Or are we compelled to cut the knot, by alluding to the inaccurate information they possessed, and by supposing some other tree of the Fir tribe to have been employed under the name of *Cedrus* for many of the purposes above indicated ; just as the early settlers in North America called the *Thuja occidentalis* by the name of the White Cedar.

Not but that in the flourishing periods of the Roman Empire the *Cedrus Libani* and *atlantica* may have been imported as timber, although not grown in the country.

The latter, however, would have come to be superseded for purposes of ornament, by another tree found in the same part of Africa, which went by the name of *Citrus*—whether from some fancied resemblance to the wood of the Lemon, does not appear. This is the tree which supplied the material for those fine tables, so much valued in ancient Rome, of which Cicero, we read, possessed one estimated at 1,000,000 *sestertia*, about £9,000, whilst still more fabulous prices are recorded as having been paid for them. The roots were espe-

[f] See Grisebach, *Flora Rumelica*, 1845.

cially prized for the rich play of colours which their variegated and knotty texture produced. But the wood was also in request for furniture, especially in veneering.

Pliny has given a long account of the qualities most esteemed in these tables, and Martial and Lucan also refer to them.

Horace[g] proposes to employ the wood, as the most precious commodity that could be selected, for a temple in which a marble statue of Venus was to be placed :—

> "Albanos prope te lacus
> Ponet marmoream sub trabe citrea;"

and Petronius Arbiter, in descanting upon the luxury of the Romans, seems to represent it as worth more than its weight in gold, where he says,—

> "Ecce Afris eruta terris
> Ponitur, ac maculis imitatur vilius aurum
> Citrea mensa."

Now I am aware that some modern writers have supposed that the wood which furnished this material was the *Cedrus atlantica;* but it is more probable that it was the *Thuya articulata,* or *Callitris quadrivalvis,* a coniferous tree, which is found at present in Algeria, and furnished those beautiful specimens of ornamental cabinet-work which were so much admired at the Exhibition in Paris a few years ago[h].

From its fragrant odour it was known in the early

---

[g] Carm. i. lib. 4. Od. 1.
[h] See *Catalogue des produits Algeriennes.* Paris, 1855.

days of Greece, Pliny says, by the name of *Thuyon*, θύον, or wood of sacrifice, under which name it is mentioned by Homer in his description of the Isle of Calypso.

The finest kind was grown, Theophrastus says, near the Temple of Jupiter Ammon.

The *Cupressus* of Pliny, and the κυπάριττος of Theophrastus, was the common Cypress of Italy and Greece, the *Cupressus sempervirens* of Linnæus.

As in modern days, it was dedicated to mourning—

> " Neque harum quas colis arborum
> Te præter invisas Cupressos
> Ulla brevem dominum sequetur."—HORACE.

Although in reality the plant is monœcous, Pliny and others describe two sexes in it, distinguishing as the female that which is more pyramidal, and by the male that whose branches are more horizontal.

There are, in fact, two varieties of Cypress recognised in modern days, viz. the pyramidal, and the horizontal, the one tapering upwards in a conical form, the other spreading out laterally [1].

Pliny regards the tree as an exotic in Italy, and states that it came originally from Crete.

Another tree allied to the Coniferous tribe, mentioned by Pliny is the *Taxus*, our Yew. He exaggerates its malignant properties, and transfers to the berry the noxious character which applies to the leaves and young shoots.

[1] Miller's Gard. Dict.

Arrows, he says, were dipped in the juice of this tree to render them more deadly, and poisons were called *toxica*, formerly *taxica*, from the name by which this tree was known.

Virgil alludes to it as communicating poisonous properties to the Corsican Honey[j], and Lucretius even represents it as destroying life by the odour of its flower, if this be indeed the tree intended :—

> "Est etiam magnis Heliconis montibus arbor,
> Floris odore hominem tetro consueta necare."

The tree is still rare in Greece, although it is mentioned in the *Flora Græca*, on the authority of Sibthorp's fellow traveller, Mr. Hawkins, as occurring on the rocks of Mount Cyllene, in Laconia.

It is, however, noticed under the name of Μῖλος in Theophrastus[k], who describes its principal properties with tolerable accuracy.

In Italy it seems to be more common, and according to Virgil[l], bows were made of it, as at present—

> "Ituræos taxi torquentur in arcus."

It thrives best, however, in colder climates, as the ancients were aware—

> "Apertos
> Bacchus amat colles, aquilonem et frigora taxi[m]."

I now proceed to certain trees recognised by ancient writers, which do not belong either to Pliny's class of *Frugiferæ*, or to those compre-

---

[j] Ec. ix. 30. [k] H. Pl. iii. 9.
[l] Georg. ii. 448. [m] Ibid., 113.

hended by him under the heads of *Glanduliferæ*
and *Piciferæ*.

And first with regard to the Lime or Linden,
which was known to the Greeks by the name of
φιλύρα, and to the Latins by that of *Tilia*.

In the former country it is rare, but Sibthorp
notices it as occurring in Laconia, and near Constantinople.

Theophrastus speaks of the male and female
φιλύρα, and Pliny also notices two varieties of the
Linden or Lime, which he distinguishes in the same
manner. These differences, however, as has been
already stated, have no reference to the sex of
the plant, but relate to the hardness and toughness
of the wood, and the thickness of its bark.

The *Tilia europæa* of modern botanists probably
comprehends both varieties, one of which is called
Tilia macrophylla, and the other *platyphylla*.

When Horace in his Odes says, " *displicent nexæ
philyrâ coronæ*," he is supposed to mean the inner
bark of the Linden ; but the φιλλυρία of Dioscorides
is identified by Sibthorp with the *Phillyrea latifolia*
of Linnæus, a shrub common in Crete, and in the
mountainous parts of Greece, and long introduced
into our gardens.

Three varieties of "*Acer*" are especially singled out
by Pliny from the many which he says exist ; nor
can there be any doubt that this term, as well as the
corresponding one Σφένδαμνος employed by Theophrastus, has reference to some members of the

Maple family; both because the Italian *acero*, and the Romaic σφένδαμι, evident corruptions of the Latin and Greek names, are applied to this tree, and also because the descriptions presented to us by classical writers correspond in a general way with the characters belonging to the Maple tribe.

The name of *Acer* is derived from the pointed character of its leaves; and Pliny describes the tree as nearly of the same size as the Lime, stating that slabs cut from it are superior even to the Citrus in the beauty of its wood, and are only less adapted for cabinet-work from their inferior size.

But in identifying the *Acer* of the Romans and the Σφένδαμνος of the Greeks with some particular species of Maple, and in imagining that the ancients applied these terms consistently to one and the same tree, modern writers have been much too precipitate.

There can be no doubt that Ovid[n] had in view a different plant, when he speaks of the *Acer vile*, even though in another passage he alludes to the variegation of colour for which its timber was prized, from that which Virgil alluded to in speaking of the wooden horse of Troy, and of the spear of King Evander, as being constructed with beams of this wood; as well as in associating the tree in question with the lofty Fir, as constituting the sacred groves of Phrygia[o].

"Nigranti piceâ trabibusque obscurus acernis."

The *Acer* of Ovid would seem to agree best with

[n] Elog. i. 11. 28.   [o] Æn. ix. 87.

the σφένδαμνος ἄγρια of Theophrastus, which that writer pronounces as useless for timber; but for the *Acer* of Virgil we must surely look to some other variety, more conspicuous for stature, strength, and durability, such as the Sycamore.

Indeed the word σφένδαμνος is said to be derived from σφενδόνη, the bezel of a ring, or the part encompassing the stone, which required to be hard and compact; and at any rate σφένδαμνινος is used adjectively in Aristophanes [p] for tough, or as we should say, heart of oak.

Now Theophrastus [q] mentions another variety which he calls Ζυγία, the wood of which, he says, is yellow, and crisp, or twisted in its fibre (οὖλον), and he adds that the people of Stagira recognised a third, affording a white and tough (εὔινον) timber, which is called κλινότροχον, a term applied to it on account of its fitness for making the rollers upon which bedsteads turn [r].

Pliny says, that the Greeks distinguish the *Acer* of the plains, which they call *glinon*, and the wood of which is white and not wavy, from that of the mountains, the wood of which is harder and more variegated. The distinction between male and female trees, which runs through all the classical writers, though in a different sense from that in which we employ it, is here noticed, and the former is said to be best adapted for ornamental purposes.

But besides these two, there is a third kind,

[p] Ach. 181.   [q] II. Pl. iii. 10.
[r] See Stapel, *Notæ in Theoph.*

called *zygia*, with a red wood, easily split, and with a pale rough bark.

He, however, as already stated, whilst noticing that there are several varieties of *Acer*, specifies only three: First, the Gallic, remarkable for the extreme whiteness of its wood, and known as the Gallic tree, a native of the countries beyond the Alps; the second, another covered with wavy spots, and from its beauty called the *pavonia*, the finest kinds of which come from Istria and Rhœtia; and the third an inferior kind, called the *crassivenum*, or thick-veined.

Now the Ζυγία of Theophrastus, the wood of which is stated to be yellow and soft, best corresponds with the *Acer platanoides*, or Norway Maple, whilst the κλινότροχος agrees better with the *Acer pseudo-platanus*, or Sycamore, as he describes it as having leaves lobed like the Plane, but less fleshy and pointed at the apex, with a bark less smooth than that of the Lime, slightly spotted, and with scanty roots, and these horizontal. Its flower, he says, is unknown, and its fruit is like that of the *Paliurus*. It grows in marshy places about Mount Ida.

Both these species of *Acer* are noticed by modern botanists as occurring in Greece, though omitted by Sibthorp. It is at least certain that only one of the four named in the *Flora Græca*, viz. the *Acer campestre*, can be identified with either of the varieties alluded to by Theophrastus, for *monspessulanum* seldom rises higher than a shrub, whilst *creticum* and *obtusifolium*, if indeed they are different species, grow only in Crete and the Archipelago.

According to this view, the *Acer* of Virgil would be the *A. pseudo-platanus*, or Sycamore of the moderns, and probably the *Acer Gallicum* is the same tree, for it is more common in France than in Italy, or in Greece, in which latter country, indeed, Sibthorp does not appear to have remarked its presence. It grows sometimes to a great height, viz. 120 feet, so that it may be compared in this respect to the Lime.

So many of the Maple tribe produce a timber, which from its variegation of colour might answer to Pliny's description of *pavonia*, that it is difficult to determine which was intended, but as the Norway, or red species, *Acer platanoides*, occurs more frequently in Greece and Italy, than our common Maple, *A. campestre*[s], it is probable that this may be the one referred to by Pliny, and also perhaps by Ovid.

The *Acer campestre*, however, often presents in its roots that beautiful veined appearance, which Pliny notices under the names of *bruscum* and *molluscum*, and which was much prized in his day for the leaves of tablets, and as a veneer for couches[t]. We have already seen that this quality is alluded to by Ovid.

Cercis siliquastrum, or Judas-tree, is abundant in Greece and Northern Italy, and Fraas supposes

it to be mentioned by Theophrastus [u] under the
name of Σημύδα. For this, however, there appears to be no satisfactory evidence, as the account given by the Greek writer seems not sufficiently descriptive.

Cytisus laburnum is mentioned by Theophrastus [x]
under the name of κολούτεα.

Pliny [y] speaks of a tree by the name *Laburnum*,
but does not describe it in a manner to identify
it with the tree now known by that name. Its
flowers, he says, which no bee will ever touch, are
a cubit in length. The latter may be intended to
designate the length of the pendulous racemes of
this tree, which is common, as he states, in the
lower parts of the Alps.

Amongst several trees belonging to the genus
Cratægus, we may mention 1st, *Cratægus Aria*, or
Mountain Ash, existing both in Italy and Greece,
and probably described by Theophrastus under the
name of ἀρία [z]. 2nd, *Torminalis*, not a plant of
Greece, but found in the south of Italy, and supposed to be noticed by Pliny [a] under that name,
although no description is given by which we can
identify it as such.

The same uncertainty exists as to the other
species of the same family, *Cratægus oxyacantha*,
our Hawthorn, which is supposed to be the *Spina
alba* of Columella, and by Dumolin, the Ῥάμνος

[u] H. Pl. iii. 14.　　　[x] Ibid., 17.　　　[y] xvi. 31.
　　　[z] H. Pl. iii. 4.　　　[a] Lib. xv. c. 23.

of Theocritus. See this further described under
that word.

Pliny [b] says, that the flowers of the *Spina* are
used for garlands, but perhaps he may refer to
the Acanthus, rather than to the May.

Cornus mascula, of Linnæus, is the Cornel of English botanists. Theophrastus describes it under the
name of κράνεια [c]. Pliny [d] mentions the flavour of
the fruit of the Cornel, *Cornum*, and in lib. xvi. 43
describes the latter as white at first, but afterwards
becoming of the colour of blood.

The *Arbutus unedo*, or Strawberry-tree of modern
botanists, was common in ancient times, as it is
at present, in Greece, and Italy. In the former
country it went by the name of κόμαρος.

Theophrastus [e] describes it as a tree not growing to large dimensions, possessing an edible fruit,
called μεμαίκυλον, a smooth bark, and a leaf intermediate between the Oak and the Bay Laurel.
Each blossom equals in size and form a long Myrtle
blossom, so that it is formed like an egg-shell cut
in half, κοῖλον ὥσπερ ὠόν ἐγκεκολαμμένον. The
fruit takes a year to ripen, so that it often is found
on the tree at the time when the new buds make
their appearance.

It is supposed to be alluded to by Lucretius [f],
but Caspar Bauhin considers the tree there spoken
of under the name of *Arbutus* to be the *Vitis idæa*,
Vaccinium myrtillus.

[b] Lib. xxi. c. 39.   [c] II. Pl. iii. 12.   [d] Lib. xv. c. 31.
      [e] II. Pl. iii. 15.   [f] Lib. v. 939.

Pliny [g] confounds the fruit of the *Arbutus* with that of the Strawberry, remarking that it is the only known instance of a similar fruit growing upon a tree, and upon the ground.

In other respects he follows Theophrastus, but admits that the fruit is not "worth eating." It was known to the Latins both by the name of *Unedo* and of *Arbutus*. Dioscorides likens the tree to the Quince.

Virgil alludes to it in Eclog. iii., and in his Georgics more than once, in the 2nd book applying to it the epithet *horrida*, which may either refer to the rough character belonging to its bark, or to the astringent nature of the tree in general.

The other species of *Arbutus*, viz. *Andrachne*, a tree peculiarly of Greece, is, however, regarded by Sibthorp as the κόμαρος of Dioscorides.

Yet Theophrastus speaks of a tree he calls ἀνδράχνη, or ἀνδράχλη, like the κόμαρος in its leaves and fruit, not of any great size, and with a bark smooth and peeling off, the latter a good characteristic of this species.

Pliny [h] says, that the *Portulaca* (our Purslane), an herbaceous plant, is called generally by the Greeks *Adrachne*, but that his *Adrachne*, or as in another place [i] he calls it, *Andrachne*, is a tree similar to the *Arbutus* in appearance, but with smaller leaves, and evergreen. He describes the bark as peeling off, as Theophrastus had done.

[g] Lib. xv. c. 28.   [h] Lib. xiii. c. 46.
[i] Lib. xxiii. c. 103.

The Flowering or Manna-bearing Ash may perhaps have been known to the ancients under the
same name, as Virgil distinguishes between the
Fraxinus and the *Ornus* in his Georgics[j].

Could the notion put forth by this Poet, of the
possibility of grafting the Pear upon the Ash, have
arisen from the profusion of white blossoms by
which the *Ornus* is distinguished?

> " Ornusque incanuit albo
> Flore pyri."

The two kinds of flowers are indeed different enough,
but the beautiful bloom of white which covers the
whole tree when in flower, might be confounded at
a distance with that of the Pear.

The property of exuding manna, however, which
belongs to the *Ornus* so common at present in
Campania, does not appear to have been observed
by the ancients.

The common Ash, *Fraxinus excelsior*, was known
to the Greeks under the name of μελία, and to the
Latins by that of *Fraxinus*.

Pliny[k] describes two varieties of *Fraxinus*, the
one long and without knots, the other short, with
a harder wood, a darker colour, and a leaf like
a laurel. Some authors suppose these differences
connected with their situation, stating that the Ash
of the plains has a spotted wood, whilst that of the
mountains is more compact.

---

[j] Lib. ii. c. 66, et seq.   [k] Lib. xvi. c. 24.

He adds several fabulous statements as to the virtues of this tree and its leaves, especially as to its being an antidote to the bite of a serpent, and as to the antipathy existing between that reptile and the plant.

Possibly Pliny's second variety may be *Fraxinus heterophylla*, a variety of the common Ash, with simple undivided leaves.

Theophrastus distinguishes the μελία and βουμελία, the latter growing in Macedonia, and of great size. Perhaps the former may be the *Ornus*, the latter the *Fraxinus*.

Pliny notices also four varieties of *Ulmus*, or Elm, the πτέλεα of the Greeks. Two of these were known, he says, to the Greeks, namely, the Mountain Elm, which is the larger of the two, and that of the plains, which is the more shrubby. To the more lofty kinds Italy gives the name of *Atinia*. It does not, he says, produce that kind of seed-vessel which we call a *samara*, and which is characteristic of the genus. This, however, is a mistake, which Columella corrects, and arose from the smaller size of the seed-vessel, which caused it to appear to casual observers to be wanting. The latter author regards the *Atinia* as synonymous to the Gallic Elm, of which Pliny makes his second variety, and he states that it is of larger dimensions than the Italian. Upon the whole, we may set it down as corresponding to our Wych Elm, or *Ulmus montana*.

The Italian Elm, the third of Pliny's varieties of

Ulmus, has its leaves lying closer together, and springing in greater numbers from a single stalk. This may perhaps be *Ulmus campestris*, our common English Elm.

The fourth variety he calls the Wild Elm, but does not describe it. It affords a curious illustration of the looseness of Pliny's classification, that he places the Elm intermediate between forest and fruit trees, because although it belongs to the former class by reason of its timber, it approaches to the latter in supporting more commonly than any other tree the branches and fruit of the Vine.

Celtis australis, or Nettle-tree, a native of the Levant, is noticed by Dioscorides under the name of λωτός.

Pliny[1] speaks of it as a tree naturalized in Italy, and the description he gives of the fruit, which, he says, is about the size of a bean, its colour that of saffron, may refer to the berries of the *Celtis*, which are represented as sweet and wholesome in the south of Europe, and indeed in modern Greek, according to Sibthorp, are called honey-berries; though it must be confessed the rest of his account applies better to the true *Lotus* of Egypt, which he describes afterwards with tolerable exactness.

He seems to have been misled by the name *Lotus*, which had been applied to the tree, and to have transferred to it the description which he had received of the plant so designated in Egypt.

[1] Lib. xiii. c. 32.

Much need not be said of the Alder and the Willow.

The former, the *Alnus glutinosa* of *L.*, the *Alnus* of the Latins, and the κλήθρος of the Greeks, is noticed by Theophrastus and Pliny as a tree planted on the borders of rivers.

Of the latter, Pliny [m] mentions several varieties, noticing their several uses for withies, wicker-work, panniers, chairs, &c., as at present. Fee and others have attempted to identify these with the varieties of Willow now recognised by botanists. The writers on agriculture, Cato, Columella, &c., point out the importance of Willow plantations, from the various uses to which the plant is applied.

We next come to the Poplar, *Populus* of the Latins, αἴγειρος of the Greeks, of which the ancients recognised only two, or at most three species.

In Homer's Odyssey we read of αἴγειρος [n], the motion of the leaves of which is compared to the rapid twinkling of the fingers of the maids of Alcinous when plying their shuttles,—

Αἱ δ' ἱστοὺς ὑφόωσι καὶ ἠλάκατα στρωφῶσιν
Ἥμεναι, οἷά τε φύλλα μακεδνῆς αἰγείροιο·

"Some ply the loom, their busy fingers move
Like poplar-leaves, when Zephyr fans the grove."

To which he adds,—

Καιροσέων δ' ὀθονέων ἀπολείβεται ὑγρὸν ἔλαιον.

"So close the work, that oil diffused in vain
Glides off innocuous, and without a stain,"—

---

[m] Lib. xvi. c. 65.   [n] Lib. vii. 106.

a line which is, as it would seem, erroneously
attributed to the tree, but which seems better to
apply to the compactness and smoothness of the
texture of the cloth, from which oil glided off.

In another passage°, Homer couples with αἴγειρος
the epithet ὑδοτρόφης.

The tallness of the tree, and its situation in
watery places, would seem to shew that the
Black Poplar was the one intended, especially as
the Aspen is not a common tree in Greece.

But Ἀχερωΐς, the lofty tree alluded to in the
Iliad[p], seems to have been the White Poplar, corresponding to the Λεύκη of Theophrastus and Dioscorides, sacred to Hercules—

"Populus Alcidæ gratissima."

Pliny notices three species of Poplar, the White,
the Black, and the Lybian.

He describes the White, as having a leaf, white
on the lower side, green on the upper. The leaves
both in this and in the Black species are rounded
when young, and partially developed, which is
the case especially with those of our White Poplar,
which in their early stage are palmated, but throw
off angular projections afterwards.

The White, he says, has a white down upon its
leaves, resembling locks of wool, in which statement
he is correct, only that the cottony matter proceeds,
not as he represents it from the leaves, but from
the seeds, which are invested in a woolly covering.

<hr>

° Lib. xvii.   ᵖ Lib. xiii. c. 389, and lib. xvi. c. 442.

He also says, that from the buds of the Poplar
exudes a resinous juice having a fragrant smell.
This exists in the buds of the White and Black Poplar, but is more remarkable in those of the *P. bal-
samifera* and *candicans*, from America, which afford
a perfume collected from the trees in the spring,
and formerly imported from Canada into England.

As to the Lybian, it is generally set down as
P. alba, though perhaps without sufficient reason.
We are, however, assured that of the genus *Populus*
there exist in Greece three species, of which one
seems really indigenous, namely, *P. alba*, with its
variety, *Populus græca*, of Aiton, found in dry situations. Next to this *P. nigra* is the most frequent,
P. tremula being the rarest of the three. The
former is found throughout the country in moist
places about brooks and springs, also in swampy
spots near the sea, as at Phalerus, and sometimes
even where there is but little water. Fraas saw it at
a height of 2,000 feet above the level of the sea.
Of all the trees which once embellished the soil of
Greece, it is the one which has best resisted the
destroying agencies that have been at work in some
parts of this country—indeed the only lofty one that
remains. These Poplars always stand in groups,
and form agreeable groves, formerly dedicated to
Hercules. They now, however, are found only in
a few localities.

Populus nigra grows further from dwelling-places,
and more rarely near rivers and ponds, although
Fraas did not observe it on any remarkable heights.

It occurs more rarely in the south, and most frequently near the Sperchius and Achelous, and in
the Olive-woods of Athens.

The rarest species is *P. tremula,* which Fraas observed only twice, namely, on the north side of
Parnes, behind Menidi, in a gorge 1,800 feet high,
and on the Achelous, behind Lithoriki. Sibthorp
also found it in Bœotia.

The Lombardy Poplar, *P. fastigiata,* the hand-
somest of the species, often regarded as indigenous
in Europe, does not appear to have been known by
the ancients.

It may have come from Persia, where it is said to
be very abundant, for before 1805 it had not made
its way even into Tuscany[q]. Into France it was
introduced in 1745, and into England about 1758.

In spite, therefore, of the opinion of Manetti[r],
who contends that it is native, because it springs
up spontaneously in places near the Po, where the
surface soil has been washed away by the inunda-
tions of the river, I think it probable that it found
its way into Italy since the time of the Romans.

The same remark applies to the Horse-chesnut,
Æsculus hippocastanum, a tree introduced into England from the Levant less than 300 years ago, and
indeed indigenous, it is said, in Northern Asia.

Yet we could hardly fix upon two trees more
completely naturalized, and entering more fully into
the landscape of Northern Europe, than the Lombardy Poplar and the Horse-chesnut.

[q] Loudon.   [r] Gard. Mag., vol. xii.

We may therefore more readily believe the report of Pliny, that the Oriental Plane came from Asia, notwithstanding its wide distribution over the southern parts of Europe.

Pliny[s] notices the Birch under the name of *Betulla*, corrupted by modern writers into *Betula*. He considers it as a native of Gaul, probably because it prefers cold and mountainous places, and was therefore first noticed in the Alps, although common in the Apennines. He describes it as remarkable for its whiteness and slender shape, and as employed for making hoops and the ribs of panniers. In Gaul they extracted from it a bitumen, as they do at present in Russia, where the oil obtained is used in preparing Russia leather.

It was regarded with a feeling of dread, in consequence of the *fasces* of the magistracy being composed of it, "as now," says Evelyn, "are the gentler rods of our tyrannical pedagogues for lighter faults." What the corresponding word in Greece may have been is somewhat doubtful, and for this reason, that the tree is not indigenous there, and therefore may not have been known to the writers in that country.

It probably, however, existed even then on the mountains, as it now does in Turkey and Asia Minor, from one of which localities came the fine specimen of *Betula alba*, v. *pontica*, which may be seen flourishing in our Botanic Garden.

[s] Lib. xvi. c. 30.

Theophrastus[t] speaks of a tree called Σημύδα, having a leaf like the Walnut, but narrower, the bark variegated, and the wood light, and only serviceable for staves. This the early commentators have chosen to identify with the Birch, but without sufficient reason. The description does not agree, and the Birch is not only not a native of Greece, but, Fraas says, will not thrive in that country when planted.

That writer, therefore, suggests it to be the Judas-tree, *Cercis siliquastrum*, but Sibthorp, more cautious, does not pretend to identify that plant with any one mentioned in ancient writings.

The Hornbeam, *Carpinus betulus*, is not common in Greece, but is noticed by Sibthorp as occurring on some of its loftier mountains, and in the neighbourhood of Constantinople. In Italy it is frequent in hilly situations, and Columella states[u] that the best handles for agricultural implements, next to the *Ilex*, are made out of Hornbeam, which is even preferable to the Ash for such purposes. It seems to have been known as *Carpinus* in Italy, but its Greek name is not ascertained.

The Hop-hornbeam, *Ostrya vulgaris*, is the *Ostrya* of the Latins, and the ὄστρυς of the Greeks.

It is alluded to by Theophrastus under the latter name[x].

---

[t] II. Pl. iii. 14.   [u] Lib. xi. c. 2.   [x] II. Pl. iii. 10.

Pliny also states that his *Ostrya* is a solitary tree, growing about rocks washed by water, and very similar in its bark and branches to the Ash [y].

Amongst the timber trees most remarkable for their size and beauty I may mention the Oriental Plane, the *Platanus* of Pliny, and the Πλάτανος of the Greeks, which though flourishing everywhere in Italy, and indeed in countries much less favoured by climate, was, Pliny says, first brought across the Ionian Sea to the Island of Diomedes, one of a small group lying off the coast of Apulia, of volcanic origin, and so nearly in a line with the volcanos of Vesuvius and Vultur on the Italian continent, that I should be tempted by its existence midway between the two countries to extend the line of igneous action existing in that quarter from Italy to Albania, and to suppose these islands to be a connecting link between the two.

The Plane-tree, Pliny says, was first planted at the tomb of Diomede, who was buried there, when, according to the fable, his companions were turned into sea-fowl, which still frequent the shrine of the hero, and seem to discriminate between Greeks and Barbarians, giving to the former a courteous welcome, but pursuing the latter with loud and clamorous cries [z].

The Plane-tree was afterwards imported from thence into Sicily, and had become in the time of

y Lib. xiii. c. 37.   z Lib. x. c. 44.

Pliny so naturalized on the Continent, that even the Morini, a nation of Belgic Gaul on the shores of the British ocean, were taxed for the privilege of enjoying its shade.

The size which these trees had attained in the time of Pliny was remarkable. In Lycia the cavity in the interior of one of them formed a species of house, 81 feet in width, and had been fitted up with seats, in which, it is added, the pro-consul of the province, Licinius Mucinus, entertained eighteen persons of his retinue at a banquet.

Dr. E. D. Clarke also describes a marvellous tree of the kind in the Island of Cos; and another in the Straits of Thermopylæ, "of unknown antiquity, self-sown in its origin, and one of many that may have flourished upon the spot ever since the Lacedæmonian soldiers were seen at the fountain combing their hair, and amusing themselves with gymnastic exercises."

Probably, however, the magnificent trees of this description now existing on the shores of the Bosphorus, called the Seven Brothers, where, it is said, Godfrey de Bouillon, with his army of Crusaders, in 1096, encamped, might surpass in beauty and interest all those instanced by the writers of antiquity.

The largest of them is 90 ft. in height, and 150 in circumference, indicating at the usual rate of growth of this tree, perhaps 1,500 years of duration.

Pliny mentions also, in the Island of Crete, an

evergreen variety of the Plane, which had even been propagated by cuttings. This statement, however, appears apocryphal.

More credit seems due to his report as to dwarf Plane-trees (*Chamæplatani*) existing in his time in Italy : for it is well known that the Chinese have long possessed the art of growing stunted varieties of the larger trees by pruning their roots, and by employing other means of arresting their development.

LECTURE III.

SHRUBS OF GREECE AND ITALY BELONGING TO THE
FOLLOWING GENERA:—

CLEMATIS — BERBERIS — CHEIRANTHUS — CAPPARIS — CISTUS—
DIANTHUS — LINUM — HYPERICUM — HIBISCUS — VITIS — RUTA
— CORIARIA — STAPHYLEA — EUONYMUS — ILEX — ZIZYPHUS —
RHAMNUS — PISTACIA — RHUS — ULEX — SPARTIUM — CYTISUS
— CERATONIA — COLUTEA — ANAGYRIS — CORONILLA — MEDI-
CAGO — PSORALEA — ANTHYLLIS — ONONIS — ASTRAGALUS —
HEDYSARUM — LOTUS — ACACIA — ROSA — RUBUS — PRUNUS —
AMYGDALUS — POTERIUM — MESPILUS — PYRUS — TAMARIX —
RIBES — BUPLEURUM — PHILADELPHUS — SEMPERVIVUM —
MYRTUS — HEDERA — SAMBUCUS — LONICERA — SCABIOSA —
ERNODEA — CONYZA — SANTOLINA — ARTEMISIA — SENECIO —
GNAPHALIUM — STÆHELINA — CENTAUREA — CINERARIA.

I NOW proceed to the identification of the shrubs
noticed in ancient writers, including all that
are mentioned by Sibthorp in his *Flora Græca* as
indigenous in Greece, and by Manetti, quoted in
Loudon's *Arboretum*, vol. iv., as now growing in
Italy, omitting from the latter list those which
are clearly of exotic origin.

The plants are named in the order in which they
occur in the Natural System.

CLEMATIS.

The eight shrubby species of Clematis now recognised by botanists as growing in Italy appear all to
have been comprehended by the Latins under the
term of Clematis, and the four noticed by Sibthorp in
Greece under that of Κληματῖτις; appellations given
to them from κλῆμα, a vine-twig, or from their

climbing habit, (κλίμαξ, a ladder). These terms, however, probably included other climbing plants, as the wild Vine is called in modern as well as in ancient Greek κλῆμα: and it is very probable that the *Periploca Græca*, a creeper common in Greece; the Bryony, ἄμπελος λεύκη, Dioscorides; and other twining plants, may be embraced under the same denomination.

M. Dumolin, in his *Flore Poétique Ancienne*, contends that the *Viburnum* of Virgil was the *Clematis viorna* of the moderns.

He remarks, that in the lines—

"Quantum lenta solent inter viburna cupressi,"

the twining character of the Clematis affords a better contrast to the erect and lofty stature of the Cypress, than does the Privet, which the *Viburnum* is commonly supposed to mean.

Ovid, too, seems to speak of the Clematis under the name of *Vitis alba :*—

"Lentior et salicis virgis et vitibus albis."

BERBERIS.

In Northern Italy this plant is common, and one species of the genus, namely, *Berberis cretica*, is met with in the Archipelago, whilst another, the *B. vulgaris*, occurs in Laconia. It is remarkable, however, that there are no traces of any description of the Berberry in Dioscorides.

Yet it is imagined by Royle[a] that the medicinal preparation called *Lycium*, or Λυκίον, which will be

[a] Trans. Linn. Soc., vol. xvii.

further noticed under the head of *Rhamnus*, may have been prepared from a species of Berberry called *B. lycium*, found abundantly at present in the mountains of Nepaul, and also, if we may believe the accounts of Arabian writers, in Syria and Asia Minor. An astringent medicine is extracted from its root and stem, and a yellow colouring-matter is also obtained from the latter, as was the case with the *Lycium* of Dioscorides.

As that writer therefore speaks of two kinds of this medicine, one the produce of Lydia and Cappadocia, and the other of India, it is very possible that the latter, which was the most prized, may have been obtained, as Royle thinks, from a *Berberis* found in Persia, and in countries still further to the East.

It is conjectured, that the Berberry may be meant by the term *Spina appendix*, a plant described by Pliny[b] as a thorn with red berries hanging from its branches, and that ὀξυάκανθα in Galen may imply the same. This word, as Sibthorp informs us, is still used in Greece to signify the Berberry. The Κολούτεα of Mount Ida, mentioned by Theophrastus[c], is by some supposed to be the *Berberis cretica;* but I shall speak of this hereafter.

CRUCIFERÆ.

Amongst the *Cruciferæ* several frutescent plants are noticed by Sibthorp in Greece, but none of them can be safely identified with modern plants.

---

[b] Lib. xxiv. c. 13.   [c] Lib. iii. c. 16.

The term Ἄλυσσον indeed is applied to some plant or other by Dioscorides, and the old botanists have chosen to adopt the same for a genus of which four suffruticose species are noticed in Greece.

But the modern *Alyssum* cannot be shewn with any degree of probability to correspond with the Ἄλυσσον of the Greeks.

Cheiranthus cheiri, indeed, may have been designated by the term λευκόϊον[d], but Sibthorp regards the *C. tricuspidatus* as better entitled to that name, especially considering its resemblance to the drawing of λευκόϊον attached to the Vienna MS.

CAPPARIS.

I may refer to my "Lectures on Roman Husbandry[e]" for some remarks on this plant, which went amongst the ancients by the same name which is at present applied to it. Royle, however, in a Paper read before the Asiatic Society, 1844, contends that it may be identified with the Hyssop of Scripture. Two species are enumerated by Sibthorp as indigenous in Greece.

CISTUS.

Eight species of *Cistus*, or Rock-Rose, including two of *Helianthemum*, are mentioned by Manetti as existing in Italy, and no less than twenty-one shrubby species by Sibthorp in Greece. Of these, the *Cistus creticus*, common in the Islands of the

[d] See "Roman Husbandry," p. 240.   [e] p. 253.

Archipelago, is identified by Sibthorp with the Λάδανον of Dioscorides, whilst the *C. villosus* is set down by him as corresponding with κίστος ἄρρην, and *C. salvifolius* with κίστος θῆλυς.

The substance called *Ledanum*, or *Labdanum*, is a resinous exudation from the *Cistus creticus*, and probably from other species of this tribe. Indeed the *C. ledon* of our gardens, commonly called the *Gum Cistus*, owes its vernacular name to the resin it exudes. Herodotus[f] tells us that Λήδανον comes from Arabia, and that it is gathered from the beards of goats, to which it is found adhering. Dioscorides[g] repeats the same story, but Pliny[h] adds, that it is obtained from the Island of Cyprus, reporting the same account as to its being collected from goats, but adding some other particulars which are evidently fictitious. In Africa and Syria, however, he says, it is procured in a different manner, namely, by passing over the plant, a string covered with wool, to which the drops of resinous matter attach themselves. A practice similar to this last is pursued, according to Tournefort, in Crete, where the gummy matter is collected by a kind of whip or rake with a double row of leathern thongs. With this the country people brush the plants, and when the whips are sufficiently charged with the juice, it is scraped off by knives and made into cakes. *Ledanum* possesses stimulant properties, and was highly esteemed in ancient times, but is now obsolete.

[f] Lib. iii. c. 112.   [g] *De M. M.* i. 128.   [h] xii. 37.

It is probable that all the species of *Cistus* known to the ancients were confounded under this same name, which in Greek was κίσθος or κίστος[i], a term nearly allied to κίσσος, which was applied to the Ivy.

DIANTHUS.

Two species of the Pink tribe run to the height of low shrubs in Greece, namely, the *Dianthus arboreus* and *fruticosus*. Theophrastus[k] speaks of a plant he calls Διὸς ἄνθος, which Neander, quoted by Atheneus, states to be used for chaplets, applying to it the epithet of εὐώδης, fragrant; but as there is no other reference in ancient writers to such a plant, modern botanists have perhaps been rather rash in identifying it with a Pink, as they have done by giving to the latter the name of *Dianthus.*

LINUM.

The common Flax, *Linum usitatissimum*, called λίνον by the Greeks, and *Linum* by the Latins, being herbaceous, does not come under our notice. And two fruticose species, mentioned by Sibthorp as occurring in Greece, namely, *arboreum* and *cæspitosum*, do not appear to have been distinguished by ancient writers.

HYPERICUM.

Six shrubby species of *Hypericum* are mentioned by Sibthorp as occurring in Greece, one of which,

---

[i] Theoph., II. Pl.   [k] Bk. vi.

H. coris, has obtained its name from being supposed to be the same which Dioscorides[1] speaks
of under the name of κόρις, a plant which he says
is also called ὑπερικόν.

Pliny[m] also speaks of two species of *Hypericum*,
one of which is distinguished as *chamæpytis*, or
Ground Pine, the other as *coris*. The description
given of both, leaves it doubtful whether our modern *Hypericum* is intended by either.

HIBISCUS

has been already alluded to in " Roman Husbandry[n]," where it is shewn that the term was
applied to various plants of the Mallow tribe, and
that it corresponded with the ἀλθαία of Dioscorides.

VITIS.

Some of the varieties of Vine described by ancient writers seem to exist at the present day—
a fact worthy of notice with reference to the much
disputed question as to the dying out of species.
Thus Pliny[o] notices a Greek vine in a manner
which would lead us to believe it meant for the
Corinth Grape or Currant of the Greek Islands.

" Græcula," he says, " non inferior Amineis boni-
tate, prætenerâ acino, et uvâ tam parvâ, ut nisi
pinguissimo solo colere non prosit."

---

[1] iii. 161.   [m] Bk. xxvi. ch. viii.   [n] p. 244.
[o] Lib. xiv. c. 4.

Columella also mentions that this variety of Vine was cultivated in several parts of Italy as well as of Greece; and Mr. Hogg [p] states, that it grows abundantly in the island of Lipari, where it is called *Passolina*.

The engraving in the Vienna edition of Dioscorides, will probably be considered as bearing more resemblance to the Currant Vine, than to the ordinary one; and Dioscorides makes mention of two varieties, one probably the common *Vitis vinifera* in its wild state, the other the *Vitis labrusca*, with a woolly leaf, the parent, as it would seem, of the Currant or Corinthian Grape.

I may add that, according to Count Odart [q], one variety of Vine, now called *Pinceau*, was known so long ago as 1394; another planted in Andalusia by the Moors still retains its characters; and that the *Cornichon* of Paris was described six centuries ago by an Arabian writer under the name of Ladies' Finger. It may be alleged, in opposition to what I have stated in page 2, that specimens of a Vine have been discovered in the Tuffs, near Montpellier, (see Planchon's late Memoir on that subject), and likewise in the brown coal of Westphalia; but whether these belong to the same species as our cultivated Vine does not appear. That the latter should have been so, is extremely improbable.

[p] Observations on some of the Classical Plants of Sicily, Hooker's Botanical Journal, 1834.

[q] *Ampelographie Universelle*, Paris, 1840.

RUTA.

Ruta graveolens is common in Greece and Italy, and *R. montana* and *chalepensis*, shrubby species, are found in the former country, as Sibthorp informs us. The generic term is noticed in "Roman Husbandry[r]," and there can be little doubt that the plant intended was the species named. It was the Πηγάνον of Dioscorides[s].

CORIARIA.

Coriaria myrtifolia, a shrub found in Italy, Spain, and other of the southern parts of Europe, does not appear to have been distinguished by the writers of antiquity.

STAPHYLEA.

Staphylea pinnata, Bladder-nut, noticed by Sibthorp, seems to be mentioned by Pliny[t] under the name of *Staphylodendron*. He says, that it is found in the countries beyond the Alps, has a wood like the White Maple, and bears a pod in which is found a kernel having the flavour of the Hazel-nut. He omits to point out the loose manner in which the capsule envelops the nut, from which it derives its English name.

EUONYMUS.

Euonymus europæus, or Spindle-tree, is found in Greece and Italy. Pliny[u] alludes to a tree of that name, and Theophrastus probably describes the same under the name of εὐώνυμον.

[r] p. 278.   [s] Sibthorp.   [t] Lib. xvi. c. 27.   [u] iii. 22.

Trees were divided by the ancients into those of good, and of bad omen. This was regarded as belonging to the former head, and hence its name.

ILEX.

The *Ilex aquifolium*, or common Holly, has been already alluded to [v]. It was called *aquifolia* by the Latins, and σμῖλαξ by the Greeks.

ZIZYPHUS.

The Jujube-tree, *Zizyphus vulgaris*, seems to be alluded to by Pliny [x] when, speaking of exotic fruits, he particularises the Zizyphus and the *Tuber*, the former of which came from Syria, the latter from Africa.

The fruit of the *Zizyphus*, he says [y], is more like a berry than an apple, although classed with the latter. Sibthorp informs us that this is the tree known in Greece at present by the name of Πα-λίουρι, which leads him to regard it as the Παλί-ουρος of Dioscorides, with which that of Theophrastus appears to agree, only that the former describes its fruit as consisting of three or four seeds included in a pod, whilst the latter designates it as a kind of berry.

The *Paliurus australis*, then, of southern Europe, *P. aculeatus* of Decandolle, *Zizyphus paliurus* of older botanists, would hardly correspond with the *Paliurus* of Pliny and Theophrastus, although the

manner in which Virgil alludes to it in his 5th
Eclogue [z], where he says that at the death of Daph-
nis, the Thistle, and the *Paliurus* with its sharp
thorns, sprang up in the place of the Violet and the
Narcissus, seems to shew, that the Jujube, a tree
prized for its fruit, could not have been there
intended.   Sibthorp identifies the *Paliurus* with
'Ράμνος τρισσός of Decandolle.

At any rate, the species which Hasselquist considered the true Christ's Thorn, and which is therefore called *Zizyphus Spina Christi*, must be distinguished from either, as having a mucilaginous Drupe
like the Jujube, although of a somewhat different
shape, and not a dry indehiscent fruit, like the
Paliurus australis.

Rhamnus.

Nine species of *Rhamnus*, or Buckthorn, occur,
according to Sibthorp, in Greece, and as many are
set down by Manetti as found in Italy.   The *R.
alaternus* is the *Alaternus* of Pliny and Columella,
the φυλίκη of Theophrastus.

Rhamnus infectorius, common in the south of
Europe, especially about Avignon, was much sought
after for its berries, which were hence called *Graines
d'Avignon*, and were used for dyeing leather yellow.

It has been supposed, that Pliny [a] alludes to this
or some other species of the same family under
the name of *Rhamnus*, from the root of which was

prepared the medicament called Lycium, which Sibthorp identifies with the Λυκίον of Dioscorides,
considering it, in his MS. notes, as derived from
the *Rhamnus infectorius*, which, he says, agrees very
well with the account given by Dioscorides.

The latter describes this plant as follows:—
" *Lycium*, which some call *Pyxacantha*, is a thorny
tree with twigs, three cubits, or even more, in
length, round which are leaves like the box, and
thick.   It has a fruit like the pepper, black, thick,
and smooth, a bark of a pale colour like a diluted
solution of its extract, and numerous oblique ligne-
ous roots.   It is chiefly found in Cappadocia, Lycia,
&c.; and loves rugged places.   A juice is extracted
from its roots, after many days' maceration with
the stem, by boiling, and then by concentrating the
liquor.   It has virtues of an astringent kind, and
is good for the eyes."

Now it is quite possible that this may refer to
a different plant from the Indian *Lycium*, which he
afterwards alludes to, and which Royle, as I have
already stated, considers to have been extracted
from a *Berberis*.

But in identifying it with the *Rhamnus* of modern
botanists, we are met with the difficulty, that the
species of this genus generally possess those purgative properties which are so manifest in our
common Buckthorn, *R. catharticus*, whilst Λυκίον
is represented as being astringent.

Hence it has been suggested, that the *Rhamnus*
of Pliny may be the *Zizyphus. Jujuba* of modern

writers, the fruit of which possesses the same astringent virtues.

But in the first place, Pliny speaks of a species of Spurge as being astringent, though he describes it at the same time as a drastic purgative, implying by the former epithet merely that its taste was rough and forbidding. And in the second place, the *Rhamnus* is mentioned by Theocritus as growing freely in Calabria, the spot in which he places the scene of his 4th Idyll; whereas we are told, that the *Jujube* was introduced into Italy from Syria three hundred years later, in the time of Augustus.

In the third place, the ʿΡάμνος of Dioscorides is stated to have been hung over the doors and windows of houses to keep out enchantments; and this very property is attributed by Ovid to the *Spina alba*, or White Thorn, which we have already identified with the Hawthorn of the present day.

Thus in the *Fasti*[b], Janus presents to the goddess Carna the White Thorn, as a means of averting any calamities from her household:—

> "Sic fatus, virgam quâ tristes pellere possit
> A foribus noxas, hæc erat alba, dedit."

And again, farther on, he says:—

> "Virgaque Jananis de spinâ ponitur albâ
> Quâ lumen thalamis parva fenestra dabat."

Now the *Zizyphus* and the *Rhamnus* do not possess white flowers, nor has any protecting influence of the kind alluded to been ascribed to either.

[b] 9th Book.

Pliny[c] speaks of the *Rhamnus* as a kind of bramble (*Rubus*), stating that they have both thorns which are not hooked, and that one is dark-coloured, the other whiter. The medicament called *Lycium*, he says, may be extracted from either plant.

Moreover, the very etymology of the word 'Ράμ-νος favours the idea that it may have been used to signify any kind of bushy shrub, as it is evidently an abbreviation of ῥάδαμνος, a young branch or sprout.

Upon the whole, then, it seems probable, as Dumolin has suggested, that the 'Ράμνος of the Greeks was often used, at least by the poets, for the Hawthorn of the present day, although it must be confessed, that the prose writers, such as Theophrastus and Dioscorides, speak of the *Rhamnus* and the *Oxyacantha* as distinct. In the old Apologue, in the 9th chapter of Judges, of the trees choosing a king, the Septuagint translates the word which we render bramble by 'Ράμνος.

There are no drawings of this plant in the Vienna MS. of Dioscorides.

PISTACIA.

Even though we were to admit Pliny's authority for the foreign origin of the Pistachio-nut, *Pistacia vera*, which a fact stated further on would lead one to dispute, other plants belonging to the natural

---

[c] Lib. xxiv. c. 76.

family of *Terebinthaceæ* must be admitted to be indigenous in the south of Europe. Amongst these
is the *P. lentiscus*, a shrub frequent in Greece, in
Sicily, and in Italy, near Nice. In the Island of
Chios this plant affords the resinous substance
called Mastic, which exudes from the bark when
incisions are made in it, and is used, as of old, for
a dentifrice, giving a sweet scent to the breath
when chewed.

Another species is the *P. terebinthus*, from which
a species of turpentine is obtained, of a more agreeable odour, and therefore more highly prized, than
the so-called Venice turpentine. Dioscorides calls it
ἔλαιον μάστιχινον.

Pliny alludes to the *P. Lentiscus* in several
places, as in lib. xii. 36, and in lib. xiv. 20; and
Theophrastus[d] seems to apply the term Σχῖνος
to the same plant, which also appears to be the
one intended by Theocritus in Idylls v. and vii.

Dioscorides also uses for the *Lentiscus* the same
name, which must not be confounded with σχοῖνος,
the *Scirpus holoschœnus* of L., according to Sibthorp;
the *Andropogon nardus* or *schœnanthus* of Fee.

Theophrastus[e] likewise mentions the *P. terebinthus*, under the names of Τερέβινθος and Τέρμινθος,
noticing the male and the female tree, and stating
that in Mount Ida and in Macedonia it is small,
whilst in Damascus and Syria it is large and fine.
In the latter country whole mountains are covered
by it.

The wood is of a glutinous character, the roots stout, and penetrating deep into the ground. The flower is like the Olive, but red. The fruit contains a resinous matter, but not in quantity sufficient to be worth collecting, the resin of the Terebinth being derived from its wood.

Pliny also alludes to it under the name of *Terebinthus*, in his 13th Book, c. 12, and 24th, c. 18, in the former of which he describes, much in the same manner as Theophrastus, the properties of the tree, and in the latter states its uses in medicine.

RHUS.

At least two species of *Rhus*, namely *Rhus coriaria* and *Rhus cotinus*, are common in Greece and Italy. *Rhus coriaria*, Sibthorp says, is the ῾Ροῦς of Dioscorides, although according to Stackhouse *R. cotinus* is the Κοκκύγρια of Theophrastus[f].

Rhus cotinus was probably not distinguished in ancient times from the former. Pliny designates it by the name of *Rhus*. It yields the Sumach of commerce, which dyes wool of an orange colour.

ULEX.

Our common Furze, *Ulex europœus*, does not seem to have been recognised by ancient writers. It appears, indeed, not to occur in Greece, and to be less common in Italy than in the north of Europe. Pliny[g] speaks of a plant called *ulex*, in appearance like Rosemary, rough and prickly,

ꜰ Lib. iii. c. 16.   ᴳ Lib. xxxiii. 21.

which, in the process of collecting gold from the
sands of rivers, is well adapted, he says, for arresting any pieces of the metal that may be carried
away during the washing, so that it was a common
practice to place a layer of it at the bottom of the
trenches along which the water was made to flow.
But it would seem that the early botanists, in identifying this plant with the Furze of the moderns,
as they have done by assigning that name to the
latter plant, have been somewhat precipitate.

Dumolin contends, that the *Ulex* of the moderns
was designated in ancient Greece by the term
ἀσπάλαθος.

Theophrastus [h] describes a plant which he calls
Σκορπίος, from the irritating character of the wound
produced by its thorns entering the flesh, as being,
except the wild Asparagus, the only one known
which is entirely destitute of leaves, it being composed of thorns, the points of which are at first
white, and afterwards become a little red. It continues in flower, he says, till after the autumnal
equinox, and is destitute of smell.

This plant Dumolin identifies with the ἀσπάλαθος
of Theocritus.

But in the first place, it seems by no means
likely that the σκορπίος was our Furze, since this
does not occur in Sicily, whilst some of the species
of *Spartium* are provided with long and much more
formidable thorns.

Secondly, I do not find that Theophrastus iden-

[h] Lib. i. 3.

tifies the σκορπίος with the Ἀσπάλαθος: on the contrary, he places the latter amongst the plants employed for perfumes.

Dioscorides mentions two kinds of Ἀσπάλαθος, one with and the other without odour; and Pliny[i] describes *Aspalathus* as a white thorn, of the size of a moderate tree, with a flower like the rose, and a root sought after for ointments; so that he must intend a very different plant.

What the *Aspalathus* of Theocritus may have been, is therefore still, I conceive, a matter of conjecture. The Poet represents it as a thorny plant, inhabiting the mountains, and couples it with the Ῥάμνοι, which we have translated, in accordance with Dumolin, as meaning briars :—

" Εἰς ὄρος ὀκχ' ἔρπεις, μὴ ἀνάλιπος ἔρχεο, Βάττε,
   Ἐν γὰρ ὄρει Ῥάμνοι τὲ καὶ Ἀσπάλαθοι κομόωντι."

' Go not to the mountain, O Battus, bare-footed,
  For on the mountain lurk the Rhamni, and the Aspalathi.'

The *Spartium villosum* and *spinosum* (or *infestum*), both occur on the dry hills of Sicily, so that I prefer upon the whole the identification of Sibthorp, who, finding that the *Spartium villosum* is called Ἀσπάλατος in Greece at the present day, considers this plant to be the Ἀσπάλαθος of Dioscorides, and would therefore probably have agreed to regard it as the one alluded to by the poet.

[i] Lib. xii. c. 52.

SPARTIUM.

Spartium was probably confounded by the ancients with *Genista*, and whilst the term *Genista*, of
which four shrubby species exist in Greece, was
employed by the Latins to designate the several
species comprehended under these two genera, that
of Σπάρτιον was used by the Greeks, or at least
by Dioscorides. Thus Virgil speaks of the *lenta
genista*[k], and Pliny[l] alludes to its use in making
withies. This would seem to refer to the *Spartium
junceum*, Spanish Broom, so much used in Spain as
a substitute for flax. The Spanish General Espar-
tero obtained his name from this plant, which his
family had maintained themselves by cultivating.

I have already pointed out the probability that
Spartium villosum may have been the ἀσπάλαθος
of the Greek writers.

CYTISUS.

The *Cytisus* of the moderns, of which Sibthorp
enumerates five shrubby species, has been generally
identified with that mentioned by the writers of
antiquity, but I have shewn in my " Lectures on
Roman Husbandry[m]" that this is a mistake, and
that the *Cytisus* of the ancients was in fact the
plant now known by the name of *Medicago arborea*.

CERATONIA.

The Carob-tree, or *Ceratonia siliqua*, has been
already noticed, p. 2. Pliny alludes to it in lib.

k Georg., ii. 12.   l Lib. xxiv. c. 40.   m p. 169.

xiii. 16 and xv. 26, under the name of *Siliqua*, adding in the former passage that the Ionians called it *Ceraunia*, and in the latter describing its pods very exactly: " Prædulces siliquæ digitorum hominis longitudo illis, et interim falcata pollicari latitudine." It has received the name of St. John's Bread, or Locust-tree, as being, according to some, the food upon which St. John the Baptist fed.

It is common everywhere in Southern Europe.

COLUTEA.

Colutea arborescens, or Bladder Senna, a shrub common in Italy, and in most parts of Southern Europe, does not appear to be noticed in ancient writers, although Theophrastus mentions a plant by the name of κολούτεα.

ANAGYRIS.

Anagyris fœtida, a plant found in Greece and Italy, is noticed by Pliny[n] under the same name. He speaks of it as a shrub with an offensive smell, a flavour like that of the Cabbage, and a seed in pods of considerable length, like a Kidney-bean. Dioscorides[o] also mentions it under the name of ἀνάγυρις and ἀνάγυρον.

CORONILLA.

Of *Coronilla* Sibthorp mentions seven species in Greece, of which only two, *emerus* and *glauca*, are fruticose. In Italy also the *Coronilla emerus* alone

---

[n] Lib. xxvii. 13.   [o] *M. M.* iii. 157.

reaches the dimensions of a shrub.   Pliny describes
it under the name of *securidaca*, and the Greeks
under that of Πελημκῖνος.

MEDICAGO.

The only shrub belonging to this genus has been
shewn to be the *Cytisus* of the ancients[p].

PSORALEA.

Psoralea bituminosa is a shrubby plant noticed
by Sibthorp in Greece.   It is probable that the
ancient Greeks confounded it with the trefoil
(τρίφυλλον), one species of which Dioscorides[q] describes as having the smell of asphalt when old, and
as therefore sometimes called ἀσφάλτιον.   Pliny[r]
speaks of it as a trefoil distinguished by the name
of *Minyanthes* or *Asphaltum*, the former from the
minuteness of the flower, the latter from the nature of its scent.

ANTHYLLIS.

Two shrubby species, *Barba Jovis* and *Hermanniæ*,
are noticed by Sibthorp in Greece, but Manetti
omits any mention of them in Italy, although the
former occurs near Nice.

Pliny[s] describes a shrub, known as *Barba Jovis*,
Jupiter's Beard, which dislikes water, is employed
in ornamental gardens, is often clipped, and has
a round bushy head with a silvery leaf.   This

---

p See " Roman Husbandry," p. 170.   q *M. M.* iii. 123.
r Lib. xxi. 30.   s Lib. xvi. 31.

silvery down which characterizes the leaf of the plant, has led to its being called *argyrophylla*, and to its identification with the former shrub, which has consequently received from botanists the name of *Anthyllis Barba Jovis*.

The *Anthyllis cretica*, called by Sibthorp *Ebenus cretica*, Theophrastus is supposed to allude to[t] under the name of Ἔβενη or Κύτισος. It appears to be a plant of Oriental origin[u].

<h2 style="text-align:center">ONONIS.</h2>

Sibthorp mentions one shrubby species as indigenous in Greece, namely, *Ononis crispa*; but it is probable that the ἄνωνις of Theophrastus[v] and of Dioscorides[x] was one of the common herbaceous species—probably the Rest-harrow, which goes by the name of ἀνωνεῖδα at present in Greece.

Pliny describes the *Ononis* as a prickly plant, having thorns on its branches, to which leaves are attached similar to those of Rue, the stem also being entirely covered with leaves, in form resembling a garland. It comes up on land newly ploughed, and is highly prejudicial to the corn, being long-lived in the extreme.

It is the *Ononis antiquorum*, the same as, or nearly allied to, our Rest-harrow, the *Ononis arvensis* of British botanists.

<h2 style="text-align:center">ASTRAGALUS.</h2>

Sibthorp mentions three shrubby species in Greece, of which one, *A. creticus*, common in

---

[t] Lib. iv. 4.   [u] Fraas.   [v] vi. 5.   [x] iii. 18.

mountainous places, is now called τράγακανθα and
τετράγκαθα. Now the former is the name applied
by Theophrastus to a plant growing in Crete, and
also in the Peloponnesus and Media, yielding the
Gum Tragacanth.

Pliny notices it under the name of *Tragacantha*,
or Goat's Thorn; the best, he says, is obtained
from Media and Achaia. From several species of
Astragalus at the present day, Gum Tragacanth is
extracted; and amongst them are the *creticus* and
aristatus of Sibthorp. Lindley pronounces the
former to be the Πότηριον of Dioscorides, the latter
his Τράγακανθα[y]. It is probable that they were
often confounded.

HEDYSARUM.

One species of shrub belonging to this genus
is mentioned by Sibthorp in Greece, namely,
H. Alhagi. It is a plant rather of the East than
of Europe, Greece being probably its most western
locality.

Another species of the same genus has been
supposed to be the *Onobrychis* of Pliny[z], but this
appears rather doubtful.

LOTUS.

This genus usually consists of herbaceous plants,
but Sibthorp mentions one shrubby species in
Greece, *Lotus Dorycnium.* It is not known by
what name this plant was distinguished in ancient

[y] Bot. Reg. 1840, Misc. p. 38.   [z] Lib. xxiv. c. 98.

Greece. The *Lotus* of the *Lotophagi*, celebrated in the 9th Book of the Odyssey,—

" Τῶν δ' ὅς τις λωτοῖο φάγοι μελιηδέα καρπὸν,
Οὐκέτ' ἀπαγγεῖλαι πάλιν ἤθελεν οὐδὲ νέεσθαι,
'Αλλ' αὐτοῦ βούλοντο μετ' ἀνδράσι Λωτοφάγοισιν
Λωτὸν ἐρεπτόμενοι μενέμεν, νόστου τε λαθέσθαι,"—

" And whoso tasted of their flowery meat
Cared not with tidings to return, but clave
Fast to that tribe, for ever fain to eat,
Reckless of home return, the tender Lotus sweet,"—

and dwelt upon by Tennyson in his beautiful poem of that name, seems to have been the *Rhamnus lotus* of Dec., called *Nebek* in Syria and Palestine, and still a favourite food amongst the Bedouins.

But with regard to the other kinds of *Lotus* alluded to in ancient writers, I have only room to refer my readers to the elaborate discussion on the subject introduced by M. Fee into his *Flore de Virgile*.

ACACIA.

In my " Lectures on Roman Husbandry[a]," I have noticed that the term *Acanthus* is sometimes applied to the *Acacia ;* a genus, however, which though noticed by Virgil, does not appear to have been ever naturalized in Italy, the Acacias now cultivated in the south of Europe being derived from the New World.

In Greece *Acacia Farnesiana* occurs, but it has been introduced in modern times.

[a] p. 241.

Rosa.

I have already alluded to the Rose in my " Lectures on Roman Husbandry[b]."

Rubus.

Sibthorp mentions four species of *Rubus* as occurring in Greece, three of which, namely, *R. idæus*, the Raspberry, *R. cæsius*, the Dewberry, and *R. fruticosus*, the common shrubby Bramble, are shrubby. They all went by the name of Βάτος, but the Raspberry was distinguished by the epithet ἰδαῖα. The same species exist in Italy, but only in mountainous situations in both countries.

Pliny[c] also mentions three varieties of *Rubus*, but one of them, which he calls *cynosbatus*, and which, he says, bears a flower like the Rose, is the Dog-rose. Another, which he says, bears mulberries, is the *Rubus fruticosus*, which yields our common blackberries ; and the third, which derives its name from Mount Ida, where it principally grows, is our Raspberry.

Dioscorides[d] speaks of the Raspberry in much the same manner.

Prunus.

Four fruticose species are mentioned by Sibthorp, but the only two which can be identified with ancient names are *Pr. spinosa*, the Sloe, and *Pr. prostrata*, a low shrub which covers the highest

[b] p. 234.   [c] Lib. xvi. c. 70.   [d] De M. M. iv. 38.

summits of Ida, Crete, Parnassus, &c. with its foliage, soon after the snow has disappeared.

The Sloe is identified by Fraas with the Σπῶδιας
of Theophrastus[e], and with the ἀγρίοκοκκυμηλέα
of Dioscorides[f]; the *P. prostrata* with the *Chamæ-
cerasus* of Pliny; but Sibthorp is more cautious in
both instances. The other two shrubby species do
not appear to have been noticed. *Pyrus domestica*
(Sib.), *Sorbus domestica* (Dec.), our Service-tree, may
have been alluded to by Theophrastus[g] under the
name of Ὄυη, of which ὄυον was the fruit.

Virgil[h] speaks of a kind of cider being made
from it :—

> " Fermento atque acidis imitantur vitea sorbis."

By Pliny, *Malus*, *Pomum*, and *Prunus* are often
used indiscriminately. See farther, under "Laurus."

AMYGDALUS.

Two dwarf Almonds, *A. nana* and *incana*, occur
in Greece, but are not pointed out by ancient
writers.

POTERIUM.

A shrubby species, *P. spinosum*, occurs on the dry
hills of Greece and the Archipelago, near the sea,
and is supposed by Sibthorp to be the Στοιβή of
Dioscorides, a name which its present vernacular
appellation, Ἀστοῖβη, confirms. The word Στοιβή
also occurs in Theophrastus, and the corresponding
one, Stöbe, in Pliny.

[e] Theoph. iii. 7.   [f] Lib. i. c. 174.
[g] Lib. iii. c. 12.   [h] Georg. iii. 380.

Mespilus.

Sibthorp enumerates four species of shrubs belonging to this genus, regarding the *M. pyracantha* as the Ὀξυάκανθα of Dioscorides. The latter name, it is well known, has been appropriated by modern botanists to our White Thorn, the *Spina alba* of the Romans, as has been above alluded to.

Pyrus.

Four frutescent species are described by Sibthorp, none of which have been identified by him, but Fraas considers that one, viz. the *P. salicifolia*, which is the commonest of all in Greece, is intended by the Ἀχράς of Theophrastus and Dioscorides, and it is now called ἀχλαδια; whilst another, the *Pyrus aria*, is conjectured to be the Ἀρία of Theophrastus[i].

Of the seven trees belonging to this genus, one, the *Pyrus communis*, our cultivated Pear-tree, was known to the Greeks under the name of ὄχνη; *P. malus*, the Apple, was the Μηλέα or Ἀγριόμηλα of Dioscorides, by which latter name it is known at the present day in Greece; *P. cydonia*, the Quince, was the Κυδώνια μῆλα of Theophrastus and Dioscorides.

Tamarix.

Tamarix Gallica is a shrub which occurs commonly in Greece and Italy, and to which Pliny[j]

[i] Lib. iii. c. 4.   [j] Lib. xiii. 37

alludes under the name of *Myrice*, stating that some persons call it *Tamarix;* but it is probable, as Fee says, that several species of Tamarisk were comprehended by the ancients under this title.

Sibthorp has identified this plant with the Μυρίκη of Dioscorides[k]; Pliny, however, states that the *Myrice* is also known as the *Erica*, and this agrees pretty well with the drawing of the plant given in the Vienna MS.

Dumolin concludes, that it was employed as a generic term for any kind of Heath, and was only applied to the Tamarisk because the ancients confounded the latter with that description of shrub.

Ribes.

Both the rough and smooth-skinned Gooseberry occur in the mountains of Greece (Sibthorp), and in addition to these the Black and Red Currant, and the *R. petræum* or Rock Currant, in Italy; but it is curious that no such plant has been noticed by the writers of antiquity, and the name itself is derived from an acidulous vegetable mentioned by the Arabian physicians, and now believed to be a kind of Rhubarb.

Bupleurum.

Two shrubby species are noticed by Sibthorp in Greece, viz. *B. fruticosum,* and *B. Sibthorpianum,* the former of which he identifies with the Σέσελι αἰθιοπικον of Dioscorides.

[k] Lib. i. c. 105.

PHILADELPHUS.

The *Philadelphus coronarius* is not noticed by Sibthorp as occurring in Greece, but it grows in the north of Italy. It is called the Mock Orange, from its scent, but is not mentioned by Pliny or Theophrastus. The φιλάδελφον spoken of by Athenæus[1] would seem to have been quite a different plant, so named from its branches inosculating so as to form one united and compact brotherhood, and thus to be well fitted for composing a stout hedge. This description certainly does not apply to the shrub referred to.

SEMPERVIVUM.

One arborescent species, *S. arboreum*, was found by Sibthorp in Cyprus, and this he has identified with the ᾿Αείζωον τὸ μέγα of Dioscorides.

A much commoner species, however, *Sempervivum tectorum*, corresponds better with the description given by ancient writers of the latter plant.

Theophrastus describes the ἀείζωον as a plant which continues ever juicy and green, has fleshy, smooth, long leaves, grows on the ground, and even on tiles—wherever, in short, the smallest quantity of soil exists for its roots to penetrate.

Dioscorides speaks of three species of ᾿Αείζωον, the largest an evergreen with fleshy leaves, at top pointed like a tongue, below concave, and with a stem a cubit in height; it is found on the mountains as well as on the roofs of houses; the second,

[1] *Deipnosophistæ,* lib. xv. c. 29.

a smaller one, found on walls and rocks, with small round pointed leaves, and pale-green flowers ; a third kind, the one called by some ἀνδράχνη, and by others Τηλέφιον, with small, thicker, and rough leaves.

Pliny[m] also describes two kinds of *Aizoum*. The larger of these grows on the roofs of houses, exceeds a cubit in height, and is somewhat thicker than the thumb ; at the extremity bearing leaves which are in shape like a tongue, fleshy, full of juice, and about as broad as a person's thumb. Some are bent downwards to the ground, whilst others stand upright, in their outline resembling an eye in shape. Hence the name *Buphthalmus* sometimes applied to it. The smaller kind grows upon walls, old rubbish, and tiled roofs. Its leaves are narrow, pointed, and juicy, the stem a palm in height.

It is evident that Pliny and Dioscorides refer to the same plant, and as the *Sempervivum arboreum* does not exist in Italy, and is rare even in Greece, the common Houseleek was the plant commonly intended under the name Ἀείζωον τὸ μέγα, although it is quite possible that the former may have been sometimes confounded with it.

Myrtus.

The Myrtle, so luxuriant in Italy, and capable of cultivation even in our northern regions, seems to have been known from the earliest times. In fact, this plant, as well as the Pistachio-nut, have both

[m] Lib. xxv. c. 102.

been found amongst the Tuffs of Mount Etna, which
are anterior to the formation of the mountain itself.
We, of course, assume, that the cultivated Myrtle
was the plant commonly known as such at present, but the wild one may · have been some
plant of the *Ruscus* family, such as our Butcher's
Broom, the *Ruscus aculeatus* of botanists, a plant of
very different affinities, but sufficiently resembling
the Myrtle in outward appearance to be confounded
with it.

Sibthorp notices the Myrtle as common in
Greece, and identifies it with the Μυρσίνη of
Dioscorides. The fruit, he says, is eaten by the
modern, as it was by the ancient Athenians.

HEDERA.

Our common Ivy abounds both in Greece and
Italy. Theophrastus[n] mentions it under the name
of κιττός, and Dioscorides under that of κισσός,
whilst Pliny[o], who gives it the name of *Edera*,
presents us with a long description of its several
kinds, which corresponds in most respects with
that given by Theophrastus.

Both specify two kinds, male and female, each of
which is subdivided into the *White*, the *Black*, and
the *Helix*. It is probable that all these are varieties
of our common Ivy, but it is difficult to identify the
descriptions given by ancient writers with any of
those recognised at present.

The *Helix*, which they represent as barren, is

[n] Hist. Pl., iii. 18.   [o] Lib. xvi. c. 62.

only the trailing variety of the plant, which in that condition bears no fruit.

The *Black* Ivy, Pliny states, sometimes bears a seed of a saffron colour, which is used for chaplets, and is known as the Ivy of Nysa, in Syria.

Now the variety with yellow berries, Royle tells us, is the one most common in the Himalayas, where it may be seen clinging to the rocks, and clasping the Oaks, so that we can readily believe the account given of Alexander in his Indian expedition having been crowned with it. Tournefort, in his "Travels in the Levant," informs us that the variety with the yellow berries is as common at Constantinople as the other kind.

In Greece and Rome it was used to decorate the *thyrsus* of Bacchus, as commemorative of the march of the god through that country.

SAMBUCUS.

Three species of Elder, or *Sambucus*, are noticed by Sibthorp as occurring in Greece, and the same number is found in Italy; *S. nigra* and *S. racemosa* being common to both; *S. ebulus* found only in Greece, and *S. laciniata* in Italy—the latter, however, is a variety of *racemosa*. Sibthorp identifies *S. ebulus*, or Dwarf Elder, with the Χαμαιάκτη of Dioscorides; and *S. nigra*, common Elder, with the ἀκτή of that author.

Pliny[p] says there are two kinds of *Sambucus*,

[p] Lib. xxiv. c. 35.

one of which grows wild, and is much smaller than the other, and is called by the Greeks *Chamœacte*, or *Helix*. In lib. xvi. c. 42, he mentions the *Sambucus* as one of the best of trees for timber ; and in lib. xv. c. 34, he alludes to the fruit of the *Sambucus* as hanging upon stalks and branches united, —the meaning of which is not very clear,—comparing it in this respect to the Ivy.

Theophrastus, however, enters more fully into a description of this plant[q], describing it as thriving especially, but not exclusively, in watery places.

This, then, is the larger kind of Elder noticed by ancient naturalists, including no doubt all the species indigenous in the south of Europe, excepting the *S. ebulus*, which is clearly pointed out by Dioscorides as a dwarf kind, resembling in its flowers and fruit the common one. To this Virgil, in his 10th Eclogue, plainly alludes, when he speaks of the *Ebulus* and its red berries :—

> " Pan, deus Arcadiæ, venit, quem vidimus ipsi
> Sanguineis ebuli baccis minioque rubentem."

LONICERA.

Although five species of Honeysuckle are noticed as occurring in Greece, and no less than nine in Italy, it is difficult to say by what names they were recognised in ancient times.

Pliny[r] describes a plant called *Clymenos*, with leaves like the Ivy, numerous branches, and a

---

[q] H. Pl., i. 6, and iii. 12.   [r] Lib. xxv. c. 33.

hollow stem, having a powerful smell and berries like those of the Ivy, growing in wild and mountainous countries. This may possibly be the Honeysuckle of modern botanists; but the Κλύμενον and Περικλύμενον of Dioscorides, Sibthorp is disposed to regard as the *Convolvulus arvensis*, and *C. sepium*, of modern botanists.

Dumolin conjectures that a plant named Αἴγιλος by Theocritus was the Honeysuckle —

Ταὶ μὲν ἐμαὶ Κύτισόν τε καὶ Αἴγιλον αἶγες ἔδοντι,
Καὶ Σχῖνον πατέοντι, καὶ ἐν Κομάροισι κέχυνται[*].

' For my goats browse upon the Cytisus and the Ægilos, and feed upon the Schinus, and lie amongst the Comari.'

This plant is also alluded to by Babrius in his Fables[t] as inhabiting the mountains.

But the chief ground for identifying it with the Honeysuckle is that both plants are regarded as a favourite food for goats, whence our English Honeysuckle is called *L. caprifolium*. This, however, seems scarcely sufficient to warrant the name being assigned to it.

SCABIOSA.

One species, viz. *S. pterocephala*, occurring in Greece, is shrubby, but we are unable to refer to it any plant named by classical writers.

ERNODEA.

The only shrub mentioned by Sibthorp in Greece as belonging to the Madder family is the *Ernodea*

[*] Idyl. v. ver. 128.   [t] Fab. iii. v. 3.

montana, which is found amongst the mountains of Crete.

It is the same as the *Asperula calabrica* of L., the *Sherardia fœtida* of Lamarck, but it cannot be identified with any classical plant.

CONYZA.

Four shrubby species are noticed by Sibthorp, one of which, *C. candida,* he suspects to be the Ἄρκτιον of Dioscorides, but the description given of the latter hardly bears out this conjecture.

SANTOLINA.

Several species of this evergreen under-shrub are mentioned in Sibthorp as occurring in Greece, and one of them, *Santolina maritima,* is supposed by Sprengel to be the *Gnaphalium chamæzelon* of Pliny [u], which he describes as having soft white leaves, and as being used as flocks for beds. Sibthorp speaks doubtfully as to its being the Γναφάλιον of Dioscorides. *Santolina chamæcyparissus* is a common shrub in Italy, often met with in our gardens under the name of Lavender Cotton.

Dumolin endeavours to identify this plant with the Πόλιον of Theophrastus [v] and Dioscorides [x], and the *Polion* of Pliny [y].

ARTEMISIA.

Of this Sibthorp mentions one frutescent species only as occurring in Greece, namely, *A. arborescens,*

---

[u] Lib. xxvii. 61.   [v] Lib. i. c. 16.
[x] Lib. iii. c. 114.   [y] Lib. xxi. c. 20, 21, 84.

the ἀρτεμισία of Dioscorides. Manetti enumerates two in Italy, viz. *Abrotanum* and *Santonica*.

Pliny[z] also specifies two, the *Santonica* and the *Pontica*, both superior, he. says, to the Italian. They were used, as at present, as a stomachic infused in wine, for which purpose Apicius introduces the plant into the materials for a banquet, recommending it to be procured from Camerina, or else from Pontus[a].

Theophrastus and Dioscorides call it ἀψίνθιον, and both represent the Pontic as the best.

SENECIO.

Sibthorp mentions one shrubby species of this genus, namely, *S. fruticulosus*, but it is not noticed by Manetti as occurring in Italy.

It is, however, improbable, that this species is alluded to in any ancient writer.

Pliny[b] says, that the plant called Ἡριγέρον by the Greeks is the *Senecio* of the Romans, both words having reference to the hoary appearance of the head of flowers when they begin to seed. It has the appearance, he says, and the softness of *Trixago* or *Chamædrys*, a plant which has been identified with the *Teucrium chamædrys* of modern botanists. It has small reddish-coloured stems, and is found growing on walls and on the tiled roofs of houses. Its name is derived from ἦρ,

---

[z] Lib. xxvii. c. 28.   [a] Dierbach, *Flora Apiciana*, Heidelberg, 1831.
[b] Lib. xxv. c. 106.

spring, and γέρων, aged, because it is white in the spring. Its head is divided into a number of downy filaments (*Spina*) protruding like a thistle; hence it is called by Callimachus *Acanthus*, and by others *Pappus*.

This description agrees very well with some of the herbaceous sorts of *Senecio*, such as our common Groundsel, except as to its turning white, or going to seed, in the spring. It must be recollected, however, that in warm climates, like Greece, it would come to seed earlier than with us.

GNAPHALIUM.

Two species of shrubby plants belonging to this genus are mentioned by Sibthorp, the best known of which is *G. stœchas*, also called *Helichrysum stœchas*, which occurs both in Greece and in Italy.

In Greece at present this plant is known by the name of ἀμάρανθον (everlasting), one of those which Dioscorides assigned to it. This author states, that it is synonymous with Ἐλίχρυσον and Χρυσάνθεμον, both expressive of the yellow colour of its petals; adding, that it is used for chaplets, has a small yellowish white stem, erect and stout, leaves scattered like the Ἀβροτόνον, a blossom (κόμην) orbicular, of a golden yellow colour, spreading out in all directions like a parasol, and bearing a resemblance to dry clusters of flowers, together with a slender root.

Theophrastus states that it has golden-coloured

flowers, a whitish leaf, a white hard stem, and a superficial thin root. The description given by Pliny I omit, as it is much the same as those above given, though less exact than that of Dioscorides.

We may, therefore, admit that *Gnaphalium stœchas* was at least one of the plants designated by the name of everlasting, and that it corresponded with the *Helichrysum* of Pliny. The latter, however, must not be confounded with the Ἐλειόχρυσος of Theophrastus[c], or with the *Heliochrysos* of Pliny[d]. Both these are wild meadow-plants, flowering in the spring, and Theophrastus associates the one he names with that species of *Anemone* which he calls *limonia*.

Hence Dumolin with reason conjectures, that it may have been the *Caltha palustris* of our meadows.

STÆHELINA.

Two frutescent species are noticed by Sibthorp, but neither can be identified with ancient names, although the plant he calls *Pteronia chamœpeuce*, which other botanists consider as a *Stœhelina*, may perhaps be the one designated by Dioscorides[e] by the name of χαμαιπεύκη.

CENTAUREA.

Amongst the twenty species of Centaury enumerated by Sibthorp, one only, viz. *C. spinosa*, is

---

[c] Lib. vi. c. 7.   [d] Lib. xxi. c. 38.   [e] Lib. iv. c. 125.

frutescent. This one, however, does not figure in
Manetti's Catalogue.

Three kinds of Centaury are mentioned by Pliny [f].
The larger, called *Chironium*, as being believed to be
the plant by which the Centaur Chiron was cured,
has been identified by the older botanists with the
Centaurea Centaurium of L., but it seems to me
quite uncertain to which particular plants this
and the other two kinds alluded to by Pliny refer.

CINERARIA.

Sibthorp mentions one shrubby species in Greece,
viz. *maritima*, but does not identify it with any
ancient name. The same also occurs in Italy, but
we are unable to point to any plant noticed by the
writers of antiquity with which it can be referred.

[f] Lib. xxv. c. 30.

LECTURE IV.

ERICA.

SIBTHORP enumerates six species of Heaths, including that now called *Calluna vulgaris*, as occurring in Greece, and the same species are common in Italy. Of these he identifies one, viz. the *herbacea* or *carnea*, with the ἐρείκη of Dioscorides, but it is probable that all the kinds of Heath observed were included by the latter under this denomination. This author describes his ἐρείκη as a shrubby or bushy tree like the Tamarisk (μυρίκη), but much smaller, remarking that the honey which bees suck from its flower is far from good.

It seems probable, as has been already stated under the head of Tamarisk, that the word Μυρίκη was often applied to the larger kinds of Heaths, such as the *Erica arborea*, and that the *Myrica* of Virgil may merely indicate Heather, although the absence of the *Ulex europæus* from Sicily (Gussone) forbids us to extend the same to the μυρίκη of Theocritus.

RHODODENDRON AND AZALEA.

The Rhododendron does not occur in Greece, but is common on the borders of the Euxine, and has been supposed to communicate those noxious properties to the honey of that country, of which Xenophon[a] speaks. Pliny[b] also seems to indicate, that it produces madness in those who partake of it: stating that this property is attributable to the flowers of a plant he calls Rhododendron, which the bees frequent.

In the preceding chapter, however, having first mentioned that at Heraclea, in Pontus, the honey is very poisonous, he goes on in the succeeding paragraph to give it as his opinion, that the poison is extracted from a plant found in that country, called "*ægolethron*," because it is fatal to beasts of burden, and to goats in particular.

The word "*Rhododendron*," however, Pliny, in lib. xvi. c. 33, uses as an equivalent to the "*Nerium*" or *Rhododaphne*, which he describes as an evergreen, bearing a near resemblance to the Rose, throwing out numerous branches from the stem, and to brutes poisonous, although to man an antidote against the venom of serpents. Now this latter

term, and the corresponding one, Νήριον or 'Ροδο-
δένδρον of Dioscorides, are identified by Sibthorp
with the *Nerium oleander*, a very common shrub
in Greece, as well as in most parts of the south
of Europe, so that Pliny must have confounded
these two plants. It has been generally supposed
that the one which rendered the honey of Pontus
noxious was the *Rhododendron ponticum*; this, however, is entirely harmless, and the shrub really
intended must have been the *Azalea pontica*, the
Chamærhododendros of Tournefort[c], which has the
reputation at present of rendering honey narcotic,
and of poisoning the goats, swine, and sheep that
browze upon its leaves[d].

VACCINIUM.

I have alluded in my "Roman Husbandry[e]" to
the probability that the *Vaccinium* of the ancients
was not the same plant as the Bilberry, or the
Vaccinium myrtillus, of the present day, but the
Hyacinth. Fee combats this opinion[f], but is
driven to suppose that the Greeks recognised two
Hyacinths, the first red, the *Lilium martagon* of L.;
the second black, which was the *Vaccinium myrtillus*. The latter, however, does not grow in watery
places, which Pliny says is the case with the *Vaccinium*. Theophrastus seems to refer to the Bilberry[g], under the name of ἄμπελος τῆς Ἴδης, but

---

[c] *Voyage au Levant*, vol. ii. p. 226.   [d] See also "Bot. Magazine,"
p. 433.   [e] p. 266.   [f] *Fl. Virg.*   [g] H. Pl. iii. 17.

Sibthorp, who notices the *V. myrtillus* as occurring in the mountains of Bithynia, does not identify it with any of the plants mentioned by Dioscorides.

LIGUSTRUM.

The *Ligustrum vulgare*, or Privet, is a common shrub both in Greece and Italy, but I have pointed out [h] that there is reason to doubt, whether the *Ligustrum* of Virgil does not mean a different plant, although it probably was intended by Pliny and Columella to indicate the one now commonly known by that name.

PHILLYREA.

Two species are noticed by Sibthorp, and four by Manetti. Of these the *latifolia*, as has been already stated, is identified with the φιλλυρέα of Dioscorides; but it is not probable that the other species were distinguished from it.

JASMINUM.

Although the common Jasmine, *Jasminum officinale*, is so abundant in most parts of Europe, and especially in Greece and the Levant, it does not appear to have been known to the ancients, and its native country is probably Arabia, from which, according to Forskahl [i], its name is derived.

CONVOLVULUS.

Three shrubby species are mentioned by Sibthorp; and one, *C. cneorum*, is noticed by Manetti.

[h] Roman Husbandry, p. 239.   [i] Fl. Ægypt. Arab., p. 59.

But it is probable that the plant alluded to by Pliny under the name of *Convolvulus*, and compared by him to a Lily, is one of the common herbaceous species, perhaps *C. sepium*, which Dioscorides says winds round trees, and in summer forms entire arbours. He calls it σμίλαξ λεία.

LITHOSPERMUM.

Two shrubby species are noticed by Sibthorp, viz. *L. fruticosum* and *hispidulum*, but in Italy they appear all to be herbaceous. Pliny[k] and Dioscorides[l] both speak of a plant called in Latin *Lithospermum*, and in Greek Λιθόσπερμον, from the hard, stony character of its seeds.

It has been suggested that the description agrees better with *Coix lachryma*, but as this is an Indian plant, it will be safer to adhere to the common opinion, that it is one or other of the species of *Lithospermum*, found in Italy and Greece, though which of them the descriptions of these writers is not sufficiently precise to enable us to pronounce.

ONOSMA.

Sibthorp mentions a shrubby species of this genus under the name of *O. fruticosa*.

The old botanists have given to this genus the name applied by Pliny[m] to a plant which he describes too concisely to admit of identification. Nor does Dioscorides[n] assist us further, in speaking of the same plant under the Greek name of Ὄνοσμα.

---

[k] Lib. xxvii. c. 74.   [l] *M. M.* iii. 148.

[m] Lib. xxvii. c. 86.   [n] *M. M.* iii. 137.

ORIGANUM.

Sibthorp only specifies two shrubby species, viz. *Tournefortii* and *Dictamnus*. The latter is the Δίκταμνος of Dioscorides, but it is probable that they were both confounded with the other kinds of *Origanum*, which appear to have had the same name in ancient times as at present, viz. Ὀρίγανον, corrupted in modern Greek to ριγανι.

Origanum majorana, a common species in Italy, seems to have been distinguished by the name *Amaracus* by the Romans, and Σάμψυχον by the Greeks. See "Roman Husbandry," p. 272, where by mistake the word *lampsana* is substituted for *sampsucum*.

TEUCRIUM.

Ten shrubby species are enumerated by Sibthorp, none of which are included in Manetti's list.

Teucrium flavum has been supposed to be alluded to by Dioscorides[o] under the name of Τευκρίον, which, he says, is also called Χαμαίδρυς, as this plant goes by the name of χαμαιδρυα at the present day.

Teucrium polium is supposed to be the Πόλιον of Dioscorides[p], and the *Polium* of Pliny[q], who describes one kind from the mountains as a shrub a cubit high, white, having a head of flowers on the top of a corymb like grey hair, with a heavy but not disagreeable smell; and the other larger, with not so powerful an odour, and inferior

---

[o] iii. 111.   [p] iii. 124.   [q] xxi. 21.

in medicinal efficacy. The smaller *Polium* may be *Teucrium montanum*; the larger one *T. polium*.

I have, however, already pointed out that Du-molin maintains that the Πόλιον of the Greeks, and the *Polium* alluded to by Pliny, was the *Santolina chamæcyparissus* of modern botanists.

PERIPLOCA.

Periploca græca is a native of Bithynia and Mount Athos, and occurs also in Italy, but it was probably confounded with other climbing plants—Clematis, Bryony, &c.—by the ancient writers, as there are no remarkable qualities by which their attention would be directed towards it.

SOLANUM.

Of the three species of Nightshade observed by Sibthorp in Greece, the *S. dulcamara* is a climber, *S. nigrum* an annual, and the third, *S. sodomeum*, a shrub. The first and second are common also in Italy.

S. sodomeum, found by Sibthorp in Sicily, of which a splendid figure is given in the *Flora Græca*, is a native of Africa and Syria. It obtained its name from being regarded as the plant which Hasselquist had identified with that bearing the famous apples of Sodom, described by Josephus and by Tacitus as fair to the eye, but when plucked, dissolving into dust and ashes[r].

---

[r] In the Book of Wisdom, x. 7, we read, "Of whose wickedness even in this day the waste land that smoketh is a testimony, and plants bearing fruit that never came to ripeness."

But for this identification, Sibthorp must be regarded as an insufficient authority, as he never visited the Holy Land, and the species commonly met with at the Dead Sea, is the *Solanum coagulans* of Linnæus, the *Sanctum* of Forskahl, and the *Hierochunticum* of Dunal, described by the latter in Decandolle's *Prodromus*. Robinson, therefore, in his " Biblical Researches [s]," appears to be mistaken in naming it the *S. melongena*, a name given by modern botanists to a whole section of the *Solaneæ*, and not to the particular species which represents the true apple of Sodom ; nor does he seem warranted in adopting the suggestion originally thrown out by Seetzen [t], that the latter plant was an *Asclepias*.

It is true, that the *Asclepias gigantea* or *procera*, the *Osher* of the Arabs, is found in a few places on the borders of the Dead Sea, and its fruit when opened contains nothing but a dusty powder. According to Robinson, too [u], it resembles a large smooth apple or orange, hanging in clusters of three or four together, and when ripe of a yellow colour ; if so, differing greatly from other members of the *Asclepias* tribe, whose seed-vessel certainly bears no sort of resemblance to either of these fruits.

But I am assured on good authority, that the *Asclepias* alluded to is an extremely rare plant in

[s] Vol. i. p. 522.

[t] See Kitto's "Physical History of the Holy Land."

[u] See Robinson's "Physical Geography of the Holy Land," 1865, p. 215.

that locality, whilst the *Solanum* is very common ; and the interior of the Apple of the *Solanum* just mentioned is likewise, it is said, frequently converted into a powder like dust, through the puncture of an insect[x].

Upon the whole, then, I am inclined to adhere to the older notion that the latter was the plant intended.

Pliny[y] says, that the *Solanum* is the same as the *Strychnon* or *Trychnon*, which latter he describes as rather a woolly shrub than a herb, with large follicles, broad and turbinated, and a good-sized berry within, ripening in November. This would seem to be the *Physalis alkekengi*.

Another kind of *Strychnon* he calls *halicacabum :* one variety of which possesses, he says, narcotic qualities, and produces death more speedily than Opium; whilst another is esteemed as an article of food.

As to the first variety, which species of *Solanum*, or of the allied genera *Atropa* and *Mandragora*, was intended, must be open to doubt, for the *Atropa belladonna*, which is a very rare plant in Greece, and not noticed by Sibthorp, seems to be mentioned by Theophrastus[z] under the name of Μαν-δραγόρας, and described as having a lofty stem like the νάρθηξ, *Ferula communis :* so that the

---

[x] The berry of the *S. sodomeum* Sibthorp describes as, "Intus viscida, amara, nauseosa, demum pulverulenta, sicca, friabilis, unde, ut videtur, nomen specificum."

[y] Lib. xxi. 105.   [z] H. Pl. vi. 2.

μανδραγόρα of this author would differ from that
of Dioscorides, which is the true Mandrake[a].

The edible variety may be one of that section
of the *Solanum* tribe, distinguished by the name
of *Melongena*, to which belongs the *Aubergine*, now
a common article of food in the southern parts
of Europe.

LYCIUM.

Two species of this climbing shrub, viz. *L. bar-
barum* and *L. europeum*, are mentioned by Sibthorp,
who identifies the latter with the Ῥάμνος of Dioscorides. This, however, does not seem consistent
with the opinion he expresses, as already stated
under the head of *Rhamnus*, that the Λυκίον of
Dioscorides was extracted from the *Rhamnus infectorius*. Although not adopting this conclusion,
I cannot identify the modern *Lycium* with the
Lycium of the ancients, as the former is wholly
destitute of any medicinal virtues.

VERBASCUM.

One species, viz. *V. spinosum*, is mentioned by
Sibthorp as occurring in mountainous places in
Crete. This, however, cannot be supposed to
have been distinguished by the ancients from the
herbaceous species which abound in Greece. To
these the term φλομος is applied in the Romaic;
and this alone would lead us to infer that the
φλόμος mentioned by Dioscorides[b], of which he

---

[a] See "Roman Husbandry," p. 274.   [b] iv. 103.

notices two kinds, white and black, was this plant. Now Pliny[c] informs us, that the φλόμος of the Greeks was the *Verbascum* of the Romans.

It was not without reason, therefore, that the old botanists gave this latter name to our common Mullein.

PHLOMIS.

Three species of *Phlomis* are mentioned by Sibthorp as occurring in Greece, but of these only one, *P. fruticosa*, is a shrub. It is identified by Sibthorp with the φλόμος ἀγρία of Dioscorides[d], which, therefore, would be a different plant from either of the two kinds mentioned above, and referred to *Verbascum*. It may be the *Phlomis* of Pliny[e], which he distinguishes from *Phlomus* above noticed.

The Roman naturalist describes two varieties, both being hairy plants with rounded leaves, and but slightly elevated above the ground. It is very doubtful, however, whether this be a true identification.

ROSMARINUS.

The common Rosemary, *Rosmarinus officinalis*, is noticed by Sibthorp as occurring in a few places in Greece, and it is met with likewise in Italy. We might expect, therefore, to see it noticed by ancient writers, and Sibthorp considers that it is the Λιβανῶτις of the ancient Greeks, as it is known

[c] Lib. xxviii. c. 13.　　　[d] Lib. iv. 104.　　　[e] Lib. xxv. 73.

by the name of Δενδρολίβανον at the present day.
Pliny describes the *Libanotis*, also, as he says, called
Rosmarinum, as having a root like to that of the
Olusatrum, and a smell nowise differing from frankincense; Virgil, Horace, and Ovid, all speak of it,
the first under the name of *Ros*, the two latter
under that of *Rosmarinum*.   Columella[f] says that
it is a good food for bees.

SATUREIA.

Five species of Savory occurring in Greece rise
to the height of a shrub, but of these one only
extends to Italy, viz. *S. montana*.   *S. Thymbra* is
identified with the *Thymbra* of Pliny, and the
Θύμβρα of Dioscorides; it is a favourite plant
with the bees, and grows on Hymettus along with
another fruticose species of the same genus, viz.
S. capitata[g], the Θύμος of Dioscorides, which is
traceable in the modern term Θυμιό or Θυμάρι.

LAVANDULA.

Three species of Lavender are noted by Sibthorp,
all three low shrubs, and in Italy one only, namely,
L. spica.   Pliny mentions a herb called *Stœchas*[h]:
describing it as an odoriferous plant, with leaves
like the Hyssop, and with a bitter taste.   Dioscorides speaks of it under the name of στοίχας,
and Sibthorp identifies this with the *L. stœchas*
which he describes.   From its abundance on the

f ix. 4.   g Sibthorp.   h Lib. xxvi. c. 27, and xxvii. 107.

islands near Hieres, in France, the latter were called by the ancients *Stœchades*. The word *Lavandula* does not occur in any Classical writer.

SALVIA.

Sibthorp enumerates no less than twenty species of *Salvia*, of which, however, only three are shrubs, viz. *S. officinalis*[i], *pomifera*, and *calycina*. In Italy the only shrubby species is *S. officinalis*. This is identified by Sibthorp with the ἐλελίσφακον of Dioscorides, described by him in the 3rd Book of his *M. M.*[j], and by Pliny[k], as possessing a powerful smell; adding, that it is considered to be the same as his *Salvia*, a plant like to Mint in appearance, white and aromatic.

I have already pointed out the probability of M. Dumolin's suggestion, that the *Salvia sclarea*, or Clary, of the moderns is the *Baccharis* of the ancients[l].

THYMUS.

Eleven frutescent species are noticed by Sibthorp, but those alluded to in ancient writers were chiefly the common herbaceous kinds, which are so abundant in the southern parts of Europe.

Thymus vulgaris seems to have been known by the name of Θύμον in Greek, and *Thymus* in Latin; *Thymus serpyllum* denoted by that of ἕρπυλλος

---

[i] Walpole's Memoirs, vol. i. p. 246.   [j] c. 38.
[k] Lib. xxii. c. 71.   [l] Roman Husb., p. 280.

by Theophrastus [m], and *Serpyllum* (*à serpendo*) by Pliny [n].

Thymus zygis, Sibth., one of his frutescent species, is so called from being supposed to be the Ζύγις of Dioscorides [o], which he distinguishes from Ἕρπυλλον by its greater size, and which, as it grows in mountainous regions, he calls Wild Thyme. *Acynos graveolens*, the *Thymus graveolens* of L., is supposed by Fraas to be the Τραγορίγανον of Dioscorides [p].

MOLUCELLA.

One frutescent species is noticed by Sibthorp, but it seems impossible to identify it with any ancient plant. It is, indeed, of rather local occurrence, namely, in Cyprus. Though not noticed by Manetti, it is met with in some parts of Italy.

PRASIUM.

The same remark applies to that genus, of which *P. majus*, noticed by Sibthorp in the Peloponnesus, the coasts of Caria, and Zante, is frutescent.

STACHYS.

Three shrubby species are mentioned as occurring in Greece, but it seems difficult to refer them to any plants noticed by the ancients, although Sibthorp has identified the commonest, viz. *S. palæstina*, with the Στάχυς of Dioscorides. The latter, indeed, is the name of a shrub men-

[m] vi. 2.   [n] xx. 90.   [o] iii. 46.   [p] iii. 35.

tioned by Dioscorides [q], resembling the Πράσιον, but larger, bearing numerous hairy leaves, hard, fragrant, white, and many shoots issuing from the root, whiter than those of the Πράσιον. It inhabits high mountains. Now the Πράσιον is regarded as a species of *Marrubium*, either *vulgare* or *creticum*. Pliny also mentions a herb called *Stachys*, which he compares to a leek, but this is probably an error, as in other respects his description, so far as it goes, agrees with that of Dioscorides. *Stachys germanica*, a very common plant in Italy, may possibly have as much claim to be regarded as the plant intended as *S. palæstina*.

VITEX.

Vitex Agnus castus is one of the commonest shrubs in Greece, and was known of old by the Greek name ἄγνος, and in earlier writers by that of Λύγος.

The name of *Agnus castus*, Pliny [r] says, was given to it from the habit of the matrons of Athens to strew their beds with it during the festival of the Thesmophora, when the strictest chastity was enjoined. He mentions two kinds, the larger, called the white, bearing a white blossom mixed with purple, the smaller with a paler, downy leaf, and a flower entirely purple.

Two varieties are also noticed by modern botanists, one having white flowers, the other purple

ones. Though in general a shrub, it sometimes
rises to the height of ten feet, and Hasselquist
observes that pilgrims make staves of it.

Salsola and Salicornia.

One shrubby species of each of these genera is
noticed by Sibthorp as occurring on the sea-coast,
near Athens, but neither of them has been identified with any plant named by classical writers.

Atriplex.

Four shrubby species are noticed by Sibthorp
in Greece, and two by Manetti in Italy.

One of them, *A. halimus*, which has established
itself on many parts of our own sea-coasts, obtained
its specific name from the ἅλιμος of Dioscorides [s],
which from his description seems to have been
meant for this shrub.

Some have regarded it as the *Batis marina*, Pliny [t],
but the description the Roman naturalist gives of
Alimon [u], seems to identify it with that plant, which
he calls a shrub, dense, white, without thorns, with
leaves like the Olive, but softer. I am not aware,
however, that the *Atriplex halimus* is ever eaten,
and therefore Fee's conjecture may be well founded,
that it is the *Atriplex portulacoides*, the young
leaves and shoots of which, preserved in vinegar,
have, he says, an agreeable taste.

[s] *M. M.* i. 120.   [t] xxi. 50.   [u] xxii. 33.

The 'Ατράφαξις of the Greek writers appears to have been the Garden Orach, *A. hortensis*, an herbaceous species.

POLYGONUM.

Two species of shrubby *Polygonum* are noticed by Sibthorp, but neither can be identified with any ancient plants.

LAURUS AND DAPHNE.

The term *Laurus* was employed by the ancients with the same degree of laxity as that in which they indulged in the case of the *Acanthus*, the *Acer*, and the like, just as that, with regard to the same term, which is admitted into the popular phraseology of the present day.

We speak of the common Laurel, the Bay, the Portugal, the Alexandrian, the Laurustinus, &c., shrubs no farther related than in the one character of being evergreen shrubs, applicable to the same uses in ornamental gardening.

In like manner Pliny enumerates the *tinus*, a plant which must have been the *Viburnum tinus*, the Laurustinus of the moderns, belonging to the family *Caprifoliaceæ*, (although some, even in his time, considered this as a tree of a separate class).

Then follows the *Royal Laurel*, sacred to Apollo, and known as the Augustan, being used in triumphs to encircle the brow of the conqueror, which is the Bay, or *Laurus nobilis* of Linnæus, belonging to the family of *Laurineæ*, and possessing something of the aroma so remarkable in certain tropical species of the same family, namely, in the Cinnamon and Cassia, plants noticed by Theophrastus and Pliny [v], but not as occurring in Europe.

The Bay Laurel itself has been regarded by some as a questionable native of Europe, and even Pliny seems to speak of it as though it were of foreign extraction, when he alludes to the care with which it was preserved, as was the case in the villa of the Cæsars [x].

But this was an exceptional case, arising from the veneration felt for that particular branch of Laurel, which, as the story goes, was held in the beak of the hen, that an eagle had let fall from a loft unhurt into the lap of the Empress Livia; and at any rate the discovery of this plant amongst the Tuffs of Castelnau, in Provence, as stated by M. Planchon in a late work on the subject, proves that it existed antecedently to man in the south of France.

The crackling which takes place when the leaves are put into the fire, arising from its oil becoming volatilized, and bursting the walls of the cells within which it had been imprisoned, was regarded by the Romans with superstitious fear, and deterred them from applying the shrub to profane purposes, or even from using it for fires at the altars. See Pliny, lib. xv. 40.

The Bay was called the barren Laurel, as the male and female flowers are on separate trees.

Sibthorp identifies it with the Δάφνη of Dioscorides, but the plant now known as Daphne is the *D. laureola*, or Spurge Laurel, which is probably the species called by Pliny *Daphnoides*.

Of *Daphne* Sibthorp mentions thirteen frutescent species, but the *D. laureola* does not occur amongst them. Fraas, however, has identified his first species, viz. *D. Tartonraira*, with the Κνέωρος ὁ λευκὸς of Theophrastus[y], both because it flowers at the autumnal equinox, and also because it has a leaf which is white and thick-skinned or hard (δερματῶδες), in contradistinction to the black kind, the leaf of which is fleshy. The former is fragrant, the latter destitute of smell.

Daphne Gnidium, the same author contends, is not the Κνέωρον μέλαν of Theophrastus, for this has neither fleshy leaves, nor fragrant leaves. He considers it the *Casia herba* of Virg.[z], now called καῦσα in Eubœa, and the Θύμελαια of Dioscorides[a], which Sibthorp also regards it. *Daphne oleoides*, or *jasminea*, Sibth., is the Χαμαίλεα of Dioscorides[b] and of Pliny. The Κνέωρον μέλαν he considers to have been the *Passerina hirsuta* of Sibthorp.

Another variety of *Laurus* is called by Pliny[c] *taxa*, and is described by him as having a small excrescence sprouting from the middle of the leaf, and forming a fringe, as it were, hanging from it.

Now this description applies so well to the *Ruscus hypoglossum* of modern botanists, that we should

----

[y] vi. 2.   [z] Georg. ii. 213; Eclog. ii. 49.   [a] iv. 170.
          [b] iv. 169.   [c] xv. 34.

be inclined to identify the two plants; but it is re-markable, that Pliny afterwards states that another so-called Laurel, namely, the Alexandrian, by some termed the Idæan, is also designated by the name of *hypoglottion*, whilst Dioscorides describes the plant he calls 'Ιδαία Ρίζα, as having leaves like the wild Myrtle, and upon them small twisted appendages (ἕλικες), from which the flower issues.

Pliny, too, in his 27th book, c. 67, speaks of the *Hypoglottion* exactly in the same terms as Dios-corides, but he just afterwards alludes to the Idæan plant, as though it were distinct from the former.

In another place Dioscorides mentions the Alex-andrian Laurel as synonymous to the *Chamædaphne*, and describes it as having its fruit placed in the middle of the leaf, a description which would apply to the *R. hypophyllum* of Linnæus, which has a flower springing from the centre of the leaf like the *R. hypoglottion*, but is destitute of that tongue-shaped bractea which is characteristic of the latter.

From the circumstance of the fruit growing from the leaf, the Alexandrian Laurel is called by Pliny *carpophyllum*.

Nevertheless others have preferred to identify the Alexandrian Laurel with the *Ruscus racemosus*, although in that species the flowers do not spring from the leaf as in the two other species.

Perhaps, if we believe that the ancients really distinguished these three species of *Ruscus*, we may be disposed to believe with Sibthorp, that the Δάφνη 'Αλεξάνδρεια of Dioscorides was the

R. hypoglottion, Χαμαιδάφνη the *R. hypophyllum,*
Μυρσίνη ἀγρια the *R. aculeatus.*

An old botanist of the sixteenth century, Fabius
Colonna, in his work entitled Φυτοβασανος, Neapoli,
1592, first pointed out the resemblance between
the *Radix Idæa* and the *Ruscus hypoglossum,* of
which latter he gives a correct drawing.

Other varieties are also mentioned, which it is
not easy to identify, but amongst them does not
occur that which is the commonest of any at the
present day, using the term Laurel in its popular
sense, namely, the *Cerasus laurocerasus,* or Laurel
Cherry, this appearing to have been unknown to
the ancients, having been introduced into Europe
from Trebizond in 1576, by Clusius, under diffi-
culties nearly as great as those stated in page 34,
with respect to the Cedar of Lebanon[d].

Osyris.

This *Osyris alba* is common in the south of
Europe, and is noticed by Sibthorp as occurring in
Greece. It is probably described by Dioscorides[e]
under the name of ὄσυρις, and by Pliny[f] under
that of *Osyris.* It has been considered the *casia* of
the Poets, but this, as we have seen, was more
probably *Daphne gnidium.*

Elæagnus.

Elæagnus angustifolia is mentioned by Sibthorp
amongst the shrubs indigenous in Greece, and

[d] See Loudon's *Arboretum,* vol. ii. p. 717.
[e] *M. M.* iv. 141.　　　　　　[f] Lib. xxvii. c. 88.

as occurring in Samos, and in Asia Minor, between Brusa and Smyrna.

Sibthorp regards it as the ἐλαία αἰθιόπικη of Dioscorides, but Fraas questions this identification.

The ἐλαιάγνος of Theophrastus was a marsh plant, and either the *Myrica gale*, or the *Salix babylonica*.

ARISTOLOCHIA.

Two shrubby species are noticed by Sibthorp in Greece, viz. *A. sempervirens* and *bætica*, one also, *A. sipho*, by Manetti, in Italy. *A. bætica* is identified by Sibthorp with the Ἀριστολοχία κληματῖτις of Dioscorides, a term which the older botanists had transferred to the herbaceous species met with in Great Britain.

Pliny[g] mentions several kinds of *Aristolochia* which had the same virtues attributed to them as the vulgar at present assign to the various kinds of Birthwort, included under the same denomination.

It does not seem possible, however, to identify the kinds mentioned by Pliny with those known in modern times; and the notices contained in Theophrastus and in Dioscorides of Ἀριστολοχία are still less precise.

EUPHORBIA.

Six shrubby species are mentioned by Sibthorp, and one by Manetti in Italy.

Pliny[h] gives a description of a plant he calls *Euphorbia*, which corresponds very well with the

general characters of the genus known by that
name; and in lib. xxvi. c. 39, he describes various
kinds of *Tithymalus*, which from the name of milk-
plant, as well as from other properties ascribed to
it, seems to be rightly identified with our *Euphorbias*.
The same word, viz. Τιθύμαλος, or Τιθυμάλλος,
occurs in Theophrastus and Dioscorides, and various
sorts are mentioned by those writers which may
with greater or less probability be referred to particular species of this genus.   But of the frutescent
species the only two that can be identified are—
1st. *E. characias*, which seems to be the Τιθύμαλος
ἀρρήν, Th., χαρακίας, Dios.; and the *E. dendroides*,
a plant introduced from the East, which is the
Τιθύμαλος δενδρώδης of Dioscorides.

Sibthorp states, that the word τιθυμαλώ is given
to the *Euphorbia* by the Greeks at present; Fraas,
however, says that he never heard that name applied to the plant during his visit in the country.

BUXUS.

Pliny notices three varieties of *Buxus*. The
Gallic, which is trained to shoot upwards in a
pyramidal form, and attains a considerable height,
is no doubt the *Buxus sempervirens*, or the common
Box of this country; but the second, which, he
says, is worthless as wood, and emits a disagreeable
odour, cannot so well be identified; and the third,
known as the Italian Box, is more spreading than
the others, and forms a thick hedge.   It is probably the dwarf variety of our common Box.

But Pliny would seem to have confounded under this name both the *Buxus sempervirens* and *balearica* of modern botanists; for whilst his general description agrees with the former, his statement that the trunks of largest size grow in Corsica would lead us to suppose that he had in view the latter.

Hawkins found the *Buxus sempervirens* on M. Pindus and in Albania, and Grisebach in Macedonia and Rumelia. It is known by the name of Πυξάρι at present, so that it is identified with the Πύξος of Theophrastus.

EPHEDRA.

One climbing shrub of this genus is noticed by Sibthorp in Greece, and by Manetti in Italy. It is common on the borders of the Mediterranean. Dioscorides describes a shrub called Τράγος[i], called also σκορπιός and τράγανον, chiefly occurring in maritime places, a palm or more in height, being a low shrub, oblong, without leaves, but with small red berries of the size of grains of wheat proceeding from its branches, pointed at top, and of an astringent taste. This is supposed by Fraas and others to be the *Ephedra distachya* of modern botanists.

The plant named *Ephedra* by Pliny[k] has been conjectured to be the same, though some regard it as another species, viz. *E. fragilis*, if this be distinct from *E. distachya*, which Grisebach doubts, and as corresponding to the Ἵππουρις of Dioscorides[l].

---

[i] Lib. iv. c. 51.   [k] Lib. xxvi. c. 20.   [l] iv. 46.

ASPARAGUS.

Four shrubby species are noticed by Sibthorp, of which the *acutifolius*, and perhaps *aphyllus*, are designated by Theophrastus [m] under the name of ἀσπάραγος. It is known in Greece at present under the name of ἀσπαράγγια. The cultivated kind of *Asparagus* does not appear to have been known in Greece, but it was much prized in Italy [n]. The wild *Asparagus* was there distinguished by the name of *Corruda*, that of *Asparagus* being confined to the kind under cultivation.

ALOE.

Aloe vulgaris, or *perfoliata*, was found by Sibthorp wild in Cyprus and Andros. It is known at present, as in ancient times, by the name of ἀλόη, but is generally met with cultivated, as the *Agave* is in Sicily. Pliny [o] confounds it with other species of Aloe from the East, and especially with that employed medicinally. It seems to have been familiarly known in Rome from the allusion to it in Juvenal [p] :—

"Plus aloes quam mellis habere."

referring to its bitter flavour.

RUSCUS.

Of this genus Sibthorp notices two shrubby species, and Manetti one.

Of these, *R. hypoglossum* has been already iden-

[m] vi. 3.   [n] Pliny, xix. 42.

[o] Lib. xxvii. 5.   [p] vi. 180.

tified in page 122, with the *Laurus taxa*, or *Alex-
andrina* of Pliny, which is the same as Δάφνη
ἀλεξάνδρεια of Theophrastus.

SMILAX.

Three climbers of this genus are noticed by Sibthorp, and one by Manetti. *Smilax aspera* is known
at the present day by the name of σμῖλαξ, so that
we need not hesitate to refer to it the plant described under the same name by Theophrastus[q],
and by Pliny[r].

The trees noticed in the preceding pages probably comprise all, or nearly all, which were recognised, or at least distinguished, by the ancients;
but the same cannot be said of the shrubs, many of
those then existing having been overlooked by the
writers whose works have come down to us.

The utilitarian character, indeed, which belongs
to all works on Natural History drawn up by
the Greeks and Romans, excluded from the consideration of their authors those natural productions which were not supposed in some way or
other to minister to man's uses or enjoyment, so
that the meagre catalogue of herbaceous plants
given in my "Lectures on Roman Husbandry"
may be easily accounted for by this circumstance;
and a similar deficiency in the shrubs enumerated
may be referred to the same cause.

[q] iii. 18.   [r] xvi. 63.

Indeed, the catalogue of either would probably have been even more scanty than is actually the case, had it not been for the medical properties ascribed by the ancients to so many more plants than modern experience justifies, owing to which various members of the vegetable kingdom have been distinguished by a place in Pliny's work, which is certainly not warranted by any real virtues belonging to them.

These, however, are often merely alluded to by their vernacular names, and therefore are not recognisable at present, whilst in other cases the descriptions given are so vague and concise, that the real nature of the plant intended is left in a great degree a matter of conjecture.

So difficult, indeed, is it to identify a modern plant from the description given by an ancient writer, that Sibthorp was glad to avail himself of two subsidiary means of determining what it might be, of which he has made a frequent use.

The first of these consisted in ascertaining the vernacular name by which the plant is known in Greece at the present day, it being presumed that the peasants retain in most instances for familiar objects the appellations handed down to them by the first settlers in the country. Thus Sibthorp, in describing his ascent of Parnassus, observes:—
"I walked out with a shepherd's boy to herbarize. My pastoral botanist surprised me not a little with his nomenclature; I traced the names of Dioscorides and Theophrastus, corrupted, indeed, in some de-

gree by pronunciation, and by the long *series an-
norum* which had elapsed since the time of those
philosophers; but many of them were unmutilated,
and their virtues faithfully handed down in the oral
tradition of the country."

And as an example of the use to which this
species of evidence may be applied, I have remarked in my "Roman Husbandry," that we have
an additional reason for believing the Misletoe of
the Oak, mentioned by Dioscorides, to have been
the *Loranthus europæus*, and not the *Viscum album*,
with which it has been usually identified, from
finding that this plant is now called ὀξὺς, a manifest corruption of ἰξὺς, whereas the parasite which
grows on the Silver Fir is the true Misletoe, and
is termed μέλλα.

The other method of identifying the plants of
Dioscorides, was by means of the drawings ap-
pended to the Vienna MS., which I have already
noticed in p. 231 of my "Roman Husbandry."

From these it may be inferred, that the word
ἄκανθος was used as a generic term for several
plants of very distinct character, agreeing only in
the circumstance of their being spinous[s]; that the
ἴον λευκόν or λευκοίον was not a violet, but some
species of cruciferous plant, and therefore perhaps,
as Sibthorp considered it, the *Cheiranthus Cheiri*, or
as it is now sometimes called, the Dame's Violet[t];
that the Ὑάκινθος was not a Larkspur, but an

<hr/>

[s] "Roman Husbandry," p. 241.   [t] p. 240.

hexandrous plant resembling the Lily, and perhaps, as Tenore conjectured, the *Gladiolus byzantinus.*

The plates also assist us in recognising Dioscorides' σίκυς ἀγρίος, which was the *Momordica elaterium,* or Squirting Cucumber, of the south of Europe; the Currant Grape, as already noticed; the plant called by Dioscorides Ἀμμωνιακὸν, which was perhaps the *Ferula orientalis,* and certainly an umbelliferous plant; the φυσαλὶς, which from the drawing would seem to have been the *Physalis alkekengi,* although Sibthorp has chosen to identify the latter with the στρύχνος ἀλικάκαβος of Dioscorides, but the plate of which bears more resemblance to the *Physalis somnifera* described by him.

I might also appeal to the engravings of the κάππαρις, or Caper plant; of the different species of μήκων, Poppy; of the ἀδίαντον, or Maiden's Hair Fern; of the Ivy, κισσὸς; of the larger House-leek, ἀειζώον τὸ μέγα; of the ἀειζώον τὸ μικρόν, the *Sedum ochroleucum* of modern botanists; of the *Aloe vulgaris,* ἀλόη; of the ἀρκευθός, or Juniper, as sufficiently resembling nature to enable us to determine at a glance the plant referred to.

In many instances, however, it must be confessed that the artist has shewn his ignorance of the object intended, by having drawn entirely from his own imagination the figure with which he has presented us.

Hence, in spite of these aids, we are often left in doubt as to the plant which the naturalists of

antiquity, in their short, confused, and even incon-
sistent descriptions, intended to bring before us.

Modern botanists, too, have increased the confusion, by appropriating to the plants they describe, without due consideration, the names found
in ancient works, and these have been adopted
without enquiry by many lexicographers.

Thus *Æsculus*, the name for a species of Oak,
is given to the Horse Chesnut, a tree which the
ancients certainly were not acquainted with; the
Cytisus, which seems to correspond with the *Medicago arborea* of modern botanists, has been transferred to the Laburnum; the Sycamore, a name
given by the Greeks to a kind of Fig, is applied
to a species of Maple, the *Acer pseudo-platanus*.

It will therefore not be a matter of surprise, that
any catalogue of trees and shrubs given by modern
writers, such as the one Loudon [u] obtained from an
Italian botanist, Signor Manetti, should be so much
more copious than that which can be collected
from any treatise of antiquity; even after excluding from the list the exotic plants introduced from
various parts of the old or new world, of which
the ancients knew nothing.

It may be remarked in general, that the number
of frutescent species is for the most part greater in
Greece than in Italy—a circumstance connected,
no doubt, with the more southern latitude of the
former country, and explicable on the Darwinian
hypothesis, by the greater mildness of the climate,

[u] *Arboretum*, vol. iv.

which allowed species, whose habit it was not to die down; to survive the winter, and thus to become perpetuated.

It will be perceived, that I have not deemed it necessary for my purpose to set down in every instance the several species belonging to each genus, as it would be hopeless to discover an ancient name appropriate to every one, so that where, in the list at the end of the volume, a Greek or Latin name is merely placed opposite to that of the genus, it is implied that the whole genus may put in a claim to the synonyme.

But although from my enumeration of the trees and shrubs now cultivated in Italy such as are exotic have been excluded, I do not adopt the same rule with regard to those named by the writers of antiquity, believing that there is often reason to doubt the authenticity of the statements given by Pliny and others as to the fact of their introduction from other regions.

Without, indeed, questioning that the fruit-trees known to the ancients had an Eastern origin, and even that some ornamental plants cultivated in their gardens and pleasure-grounds may have been derived from other countries, I am inclined to believe, with the younger Decandolle, that wherever a particular kind of tree has established itself over a wide extent of country so as to constitute a forest, it ought, unless the contrary be proved, to be regarded as natural to the soil.

Now this seems to hold good with the Chesnut

at least, if not with the Walnut; the former covering in Tuscany and the south of France large tracts
of country, and seeming as much entitled to the term
of *aborigines* as the Oak or the Fir.

Nor can it be said that these trees require the
fostering care of man to maintain themselves in the
countries where they existed in ancient days, as the
Chesnut seems to do in England, and the Date
Palm in Italy and Greece, for both ripen their fruit
to perfection, issuing spontaneously from the ground
from seeds self-planted, and are able to withstand
the most rigorous cold ever experienced in those
countries without being dwarfed or blighted in
consequence.

This position does not of course exclude any
speculations that may be indulged in, as to the
mode in which plants became disseminated by
natural causes from the centres in which each may
be supposed to have been originally created, under
a different configuration of sea and land than that
which now exists.

But this was at least a process requiring a vast
duration of time, probably indeed being long antecedent to the peopling of the country by its human
inhabitants.

It seems certain, that forests of Oak, of Chesnut,
and of Beech must have established themselves
throughout Europe before man took possession of
the country, for unless we conceive that the first
settlers brought with them seed-corn and other domestic vegetables, and had already passed through

the previous stages of hunter and pastoral life, we must adopt the early traditions which represent them as living upon acorns, beech-nuts, and chesnuts.

Thus Lucretius :—

> " Glandiferas inter curabant corpora quercus
> Plerumque; et, quæ nunc hiberno tempore cernis,
> Arbuta puniceo fieri matura colore,
> Plurima tum tellus, etiam majora ferebat ;
> Multaque præterea novitas tum florida mundi
> Pabula dia tulit, miseris mortalibus ampla ʳ."

> " But acorn-meals chief culled they from the sheds
> Of forest oaks; and in their wintry months,
> The wide wood-whortle with its purple fruit
> Fed them, then larger and more amply poured,
> And many a boon besides, now long extinct,
> The fresh-formed earth her hapless offspring dealt."

As Decandolle observes, the mere conveyance of the seeds of amentaceous and coniferous plants across an arm of the sea by natural causes is almost inconceivable, and the spontaneous establishment of a forest of such trees absolutely impossible, unless man took the trouble of bringing it about; whilst the extension of such forests in the early ages of Greece and Rome would seem to throw back their antiquity to a date, when the human race was too rude and unsettled to have attempted such an undertaking.

Chesnuts then, as well as the other *Cupuliferæ* which are found in forests throughout Europe, may fairly be regarded, in spite of the authority of

ʳ Lib. v. 937—942.

Pliny, as indigenous to the countries in which they are found.

If the gaps which occur in the continuity of these forests create a difficulty, if it be objected that large intermediate spaces exist where such trees are entirely wanting, it is much more easy to conceive that they had died off in the latter localities, than that they had been planted in the countries where they are found through the instrumentality of man.

Thus certain species of Oak seem to be undergoing diminution at the present time. *Q. cerris*, spread over the whole of Asia Minor, is now found in Europe only in a few isolated spots, as in the Apennines, in Sicily, near Besançon, and in the west of France along the Loire.

The *Pinus excelsa*, so abundant in the Himalayas, has been detected, as we have seen [x], by Grisebach on the mountains of Rumelia, but is not known to occur in any intermediate position. These may be regarded as *oases* in the midst of vast spaces over which the species is unknown. The common Oak even seems to shew symptoms of wearing out, not establishing itself spontaneously in countries where it has been exterminated, and, where it exists, suffering from the destruction of its forests through the agency of man and beasts. On the other hand the Beech appears to be extending itself throughout Europe, owing partly to drainage, partly to a fit soil being prepared for it by the *detritus*

[x] See p. 32.

of leaves of other trees, and partly from its stifling other plants by its own foliage, and thus obtaining exclusive possession of the soil.

Of the extinction of trees within periods which, although very remote, must in a geological sense be regarded as modern, we have a striking instance in the fact stated by Danish naturalists, and recorded in Lyell's "Antiquity of Man," p. 9, that although the Scotch fir is not a native of the Danish Islands, and when introduced there does not appear to thrive, yet that its trunks are met with in the bogs at various depths, associated with flint implements.

In the same bogs, but at a higher level, are found prostrate trunks of the sessile variety of the common Oak, and still higher up some of the pedunculated variety of the same tree, together with the Alder, Birch, and Hazel. Now the Oak in later times has been almost superseded throughout Denmark by the Beech.

Other trees, such as the White Birch, characterise the lower part of the bogs, but disappear from the higher; while others, again, such as the Aspen, *Populus tremula*, occur at all levels.

This, and other corresponding facts that might be cited from the animal kingdom, shew, that even under the present conditions of the earth's surface particular plants disappear, and others take their place.

I have alluded to this fact in a paper read before the Natural History Section of the British Association, at the meeting held in Bath [r], and it may not be out of place to recapitulate here the arguments which have led me to suspect, that even when the external conditions remain unchanged, each species, like every individual belonging to it, has its days numbered, and that the period assigned to its duration may be extended indeed by favourable, and abridged by unfavourable external conditions, but in no case can transcend certain definite limits.

I there remarked, that we seem in this instance to trace the workings of two antagonistic principles; the first, that which aims at handing down to the offspring the leading characteristics of its parents; the second, one which causes the vigour of the race gradually to decline, and its peculiar excellences to be effaced, owing to the balance, upon which the harmonious workings of the system depend, being destroyed, through the undue preponderance of one element, and the diminution or loss of another.

In both instances, however, nature seems to have provided means for postponing this inevitable termination for a longer or shorter interval of time; namely, by those variations from the primitive type, which are to a certain extent brought about by the mere process of sexual reproduction, and which are still further secured by those contrivances for preventing self-fertilization, to which Mr. Darwin and others have of late called particular attention.

[r] This paper is published *in extenso* in the "Gardener's Chronicle" for October, 1864.

In animals the frequent union of individuals too nearly identical is checked by the power of locomotion which they possess, and a still further tendency to variation is brought about by the changes in climate, food, &c. which they have to encounter.

But in plants, special contrivances against self-fertilization appear to have been required for preventing the too rapid deterioration of the race, so that, even in cases where the male and female organs grow together, it has been provided, that the pollen of one flower should be conveyed in various ways to the stigmas of another; and it seems a significant fact, that in so much the greater majority of instances, trees, and other plants of long duration and of vigorous growth, should possess either monœcous or diœcous flowers, as if it were intended by this arrangement to renovate more effectually the vitality of the plant, and thus to secure to the species a longer period of existence.

Yet with all these provisions for prolonging the life of a species, its days, like those of the individuals composing it, are numbered, and the only question that remains for us to consider is, whether its dying out is to be regarded simply as the result of the altered condition of climate, soil, &c. to which it has been subjected, or occurs from some inherent tendency to decay in its own organization.

The former explanation is the most obvious one, and may prove satisfactory to certain minds; for the gradual sinking of temperature which has taken place in the crust of the globe, down to the glacial period, and its subsequent elevation during that in which we live, suggest causes for the disappearance of certain species, and for the substitution of others, which may be deemed sufficient to afford an adequate solution of the problem.

Nevertheless, if these effects are exclusively due to climate, certain other conditions, at least, besides that of temperature, must be concerned in producing them.

Take, for instance, the case of the *Wellingtonia.*

This tree, or one nearly allied to it, existed generally throughout Europe during the Miocene period.

The *Lignites* of Bovey Tracey, in Devonshire, are supposed to be principally made up of its remains, and an allied species is widely distributed over Europe and America, from Italy to Greenland, during the Miocene period, when the climate is supposed to have been almost sub-tropical.

At present, its only native *habitat* is California, yet it does not seem to have been limited to this one region by the gradually decreasing temperature, since in that case it would not have survived the glacial period, nor would it prove capable of resisting the present winters of Great Britain.

Mr. Bateman, who has introduced this tree largely into his plantations in Cheshire, considers it hardier than the *Deodara;* yet from some mysterious cause unconnected with temperature, all his specimens were blighted in the spring of 1864, whilst the

Deodaras did not suffer. Does not this fact suggest some unexplained condition of climate which, operating upon the species over a long lapse of years, has at length reduced it to the narrow limits in which it is found native—unless indeed we fall back upon that higher law, which prescribes to each species a limited period of duration, and supposes this period to be arriving in the case of this remarkable vegetable production?

It may be said, indeed, that such an inference is contradicted by the vigorous growth and gigantic proportions which this tree assumes in the spots where it still exists. Yet the same remark would apply to several other trees of great size and vigour, such as the *Dracæna draco*, the *Callitris quadrivalvis*, and the *Glyptostrobus heterophyllus*[z], now confined to one country, though formerly of much more extensive distribution.

In these cases, also, it is difficult to point out what changes of climate could have led to their being thus restricted within their present range.

But it is time to bring these Lectures to a termination, in concluding which I cannot help remarking upon the small progress made in natural knowledge between the period of Alexander and Trajan, a distance of time amounting to not less

---

[z] *Callitris*, found now in Algeria, has been detected in the Miocene formation of Aix, in Provence. (Count Saporta in the *Ann. des Sc. Nat.*) *Glyptostrobus*, found now in China and Japan, had spread during the tertiary period over Switzerland; *Dracæna draco* is now confined to Teneriffe.

than from four to five hundred years. In accuracy of information Theophrastus and Aristotle both greatly exceed Pliny, whose work, although invaluable as a Cyclopædia, bears evident marks of being a compilation, and not the result of original research.

The Romans, indeed, seemed to have acted towards the Greeks, as our mediæval writers did towards the ancients, and instead of observing for themselves, were contented with copying from preceding authors, whose statements had with them the force and authority of ocular demonstration.

Thus there is often a remarkable similarity between the descriptions of plants given by Pliny and Dioscorides, shewing either that one copied from the other, or that both derived their information from some common source. Yet neither writer ever alludes to the other, unless Pliny may be supposed to do so in a passage of his 36th Book, c. 37, where, after describing a stone called Schistos, possessing medicinal virtues in diseases of the eye, much in the same terms as Dioscorides does in his 5th Book, c. 145, he adds, " Hæc est sententia eorum, qui nuperrimè scripsere." Pliny, indeed, is very copious in his citations of antecedent authors, but perhaps at the time when he drew up his list, Dioscorides had not risen into that celebrity which he afterwards obtained, as being the writer who has handed down to us the fullest account of the simples employed by the ancients.

Nevertheless it seems to be pretty well ascertained, that Dioscorides lived in the age of Pliny, for he dedicates his books to one Areus Asclepia-deus, who was the friend of Læcanius Bassus, one of the Consuls in A.D. 64, during the reign of Nero, fifteen years before the great eruption of Vesuvius in which Pliny perished.

At all events, it seems probable that both these writers derived much of their information from one common source, and it is to be feared that, in too many instances, this was obtained not directly from the observation of Nature, but from earlier autho-rities, such as Theophrastus and others of inferior credibility.

Another remarkable proof of their practice of appealing servilely to authority for the facts they record, in preference to going themselves to the fountain-head, is afforded by the drawings of plants which accompany some of the MSS. of Dioscorides, the oldest of which extant, is that of Vienna, already alluded to in my Lectures[a], having been executed in the fifth century, and therefore not very far dis-tant from the age in which Dioscorides himself flourished; yet they bear on the face of them the appearance of being copies, and often blundering copies, of drawings of even an earlier date.

And a MS. several centuries more recent than this, which had been originally procured from the East by a Florentine nobleman named Renuccini, in whose family it had remained for more than a century, and which is now in the possession of

[a] "Rom. Husb.," p. 231.

Sir Thomas Phillips, of Middle Hill, in Worcester-
shire, was likewise accompanied with coloured drawings, in many cases identical with those in the
earlier MS. alluded to.

The works of Columella, too, are in most respects an amplification in more elegant Latinity of
the earlier writings of Cato and Varro, and very
possibly the two latter would have been found to
be taken from the great Carthaginian work on
Agriculture by Mago, if the latter had come down
to us.

How mortifying it is to think, that whilst these
repetitions of facts, and even of old fables, recorded
by many of the authors referred to, might have
been so well spared, we should have to deplore
such gaps in the history and literature of antiquity,
as have arisen from the loss of many of the Books
of Livy, and from the almost entire destruction of
the Comedies of Menander and Epicharmus.

CATALOGUE

OF

TREES AND SHRUBS INDIGENOUS IN GREECE AND ITALY,

WITH THEIR ANCIENT GREEK AND LATIN SYNONYMES.

N.B. *Rom. Husb.* denotes my Lectures on Roman Husbandry, 1857; *Th.*, Theophrastus; *Dios.*, Dioscorides; *Pl.*, Pliny; *Virg.*, Virgil; *Ov.*, Ovid; *Sib.*, Sibthorp; *Dec.*, Decandolle; *Linn.*, Linnæus. Under each genus those species only are enumerated, which can be identified with some degree of probability with the plants named in Greek or Latin writers.

Modern Greece.	Modern Italy.	Ancient Greek Names.	Ancient Latin Names.	Page
Clematis, 4 *species* .	8 *species* .	Κληματῖτις .	Clematis(Pl.), Viburnum (Virg.) . .	64
Berberis, 2 *sp.*	2 *sp.* . .	. . . . .	Spina appendix . . .	65
—— cretica .	. . . .	Κυλούτεα(Th.)?		
Capparis, 2 *sp.* .	1 *sp.* . .	Κάππαρις . .	Capparis . .	67, and Rom. Husb., 253
Dianthus, 2 *sp.* . .	1 *sp.* . .	Διός ἄνθος ? .	. . . . .	69
Cistus, 21 *sp.* . .	6 *sp.* . .	. . . . .	Leda . . .	67
—— creticus . .	. . . .	Λάδανον		
—— villosus . .	. . . .	Κίστος ἄρρην		
—— salvifolius . .	. . . .	Κίστος θῆλυς		
	Helianthemum, 2 *sp.*			
Linum, 2 *sp.* . .	. . . .	Λίνον . . .	Linum . .	69
Hypericum, 6 *sp.* .	3 *sp.* . .	Κόρις? or Ὑπερικόν?	Hypericum ?	69
Hibiscus, 1 *sp.* . .	1 *sp.* . .	Ἀλθαία . .	. . . .	70
Ruta, 3 *sp.* . . .	1 *sp.* . .	Πηγάνον . .	Ruta . . .	72
Coriaria, 1 *sp.* . .	1 *sp.* . .	. . . . .	. . . .	72
Staphylea, 1 *sp.* . .	1 *sp.* . .	. . . . .	Staphylodendron . .	72
Euonymus, 1 *sp.* . .	3 *sp.* . .	Εὐώνιμον? .	Euonymus ? .	72
Ilex, 1 *sp.*	3 *sp.* . .	. . . . .	. . . .	13
—— aquifolium .	. . . .	Σμῖλαξ . . .	Aquifolia .	73, and Rom Husb., 270

L

Modern Greece.	Modern Italy.	Ancient Greek Names.	Ancient Latin Names.	Page
Vitis, 1 *sp.* . . .	3 *sp.* . . .	Ἄμπελος ἀγρία	Vitis . . .	70
Rhamnus, 9 *sp.* . .	9 *sp.* . . .	. . . . .	. . . . .	74
Zizyphus, 2 *sp.*				
—— vulgaris . . .	. . . . .	Παλίουρος (Dios.&Th.)	Zyzipha (Jujubarum arbor.)	
—— paliurus (Paliurus australis)	. . . . .	Ῥάμνος τρισσός (Dios.)	Paliurus (Virg.)	74
Pistacia, 2 *sp.* . .	. . . . .	. . . . .	. . . . .	77
—— lentiscus . .	. . . . .	Σχῖνος . .ˋ.	Lentiscus	
—— terebinthus . .	. . . . .	Τερέβινθος .	Terebinthus	
Rhus, 2 *sp.* . . .	6 *sp.* . .	Ῥοῦς, or Κοκκύγρια	Rhus . . .	79
Acer, 4 *sp.* . . .	10 *sp.* . .	Σφένδαμνος .	Acer . . .	44
—— pseudo-platanus	. . . . .	Κλινότροχος .	Acer gallicum (Pl.), Acer (Virg.)	
—— platanoides ⎰ —— campestre ⎱	. . . .	Ζυγία . . .	⎰ Pavonia(Pl.) ⎱ Acer (Ov.)	
Tilia, 1 *sp.* . . .	1 *sp.* . .	Φιλύρα . .	Tilia . . .	44
Cercis, 1 *sp.* . . .	1 *sp.* . .	Σημύδα ? . .	. . . . .	48
	Ulex, 1 *sp.* .			
Spartium, 3 *sp.* . .	1 *sp.* . . .	. . . . .	Genista . .	80
—— villosum .	. . . . .	Ἀσπάλαθος		
	Genista, 4 *sp.*	. . . . .	Genista . .	82
Cytisus, 5 *sp.* . . .	5 *sp.* . .	. . . . .	. . . . .	82, and Rom. Husb., 169
	Ceratonia, 1 *sp.*	. . . .	Siliquæ prædulces, or Ceratonia	82
Colutea, 1 *sp.* . . .	1 *sp.* . .	Κολόυτεα		
Anagyris, 1 *sp.* . .	1 *sp.* . .	Ἀνάγυρις . .	Anagyris . .	83
Coronilla, 2 *sp.* . .	1 *sp.* . .	. . . . .	. . . . .	83
—— emerus . .	. . . . .	Πέληκῖνος .	Securidaca	
Medicago, 1 *sp.* . .	1 *sp.* . .	Κύτισσος . .	Cytisus . .	82, and Rom. Husb., 170
Anthyllis, 2 *sp.* . .	. . . . .	. . . . .	. . . . .	84
—— barba Jovis .	. . . . .	. . . . .	Barba Jovis .	
Ononis, 1 *sp.* . . .	. . . . .	Ἄνωνις, or Ὄνωνις	Ononis . .	85
Astragalus, 3 *sp.* . .	1 *sp.* .			
—— aristatus .	. . . . .	Τραγάκανθα .	Tragacantha .	85
—— creticus . .	. . . . .	Πύτηριον . .	. . . . .	
Hedysarum, 1 *sp.* .	. . . . .	. . . . .	. . . . .	86
Ebenus, 1 *sp.* . .	. . . . .	Ἔβενη, or Κύτισος . .	. . . . .	85
Lotus, 1 *sp.* . . .	. . . . .	. . . . .	. . . . .	86

Modern Greece.	M dern Italy.	Ancient Greek Names.	Ancient Latin Names.	Page
Psoralea, 1 *sp.*		Ἀσφαλτιον	Minyanthes, or Asphaltum	84
Prunus, 6 *sp.*	2 *sp.*			2 and 88
—— spinosa		Σποδίας (Th.), Κοκκυμηλία (Dios.)	Prunus	
—— prostrata		Χαμαικεράσος	Chamæcerasus	
—— domestica (sorbus domestica, Linn.)		Ούη (Th.)	Sorbum	
Amygdalus, 3 *sp.*	4 *sp.*			6
—— communis		Ἀμυγδάλη	Amygdalus	
	Persica, 2 *sp.*		Malus Persica	2
	Armeniaca, 1 *sp.*		Armeniaca	2
Cerasus, 1 *sp.*	Cerasus, 5 *sp.*	Κερασία	Cerasus	2
Poterium, 1 *sp.*		Στοιβὴ	Stöbe	89
	Spiræa, 1 *sp.*			
Rosa, 6 *sp.*	18 *sp.*	Ῥόδον	Rosa	88
Pyrus, 10 *sp.*	6 *sp.*	Ὄχνη (Th.)	Pyrus	2 aud 90
—— communis		Ἀχράς (Dios.)	Pyrus	
—— malus		Ἀγριόμηλα	Malus	
—— cydonia		Κυδώνια μῆλα	Malus Cydonia	
Cratægus (mespilus, Sib.), 6 *sp.*	6 *sp.*			4 and 90
—— aria		Ἀρία		
	Torminalis		Torminalis	
—— oxyacantha			Spina appendix, Spina alba	49
Cornus, 1 *sp.*	2 *sp.*			
—— mascula		Κράνεια	Cornum	50
Fraxinus, 1 *sp.*	2 *sp.*			
—— excelsior		Βουμελία (Th.)	Fraxinus	52
Ornus, 1 *sp.*	9 *sp.*			
—— europæa		Μελία (Th.)	Ornus	52
Tamarix	1 *sp.*	Μυρίκη	Myrice	90
Ribes, 2 *sp.*	4 *sp.*			91
	Philadelphus, 2 *sp.*			92
Myrtus	1 *sp.*	Μυρσίνη	Myrtus	93
Punica, 1 *sp.*	2 *sp.*	Ῥόα	Malum punicum	2
Sempervivum, 1 *sp.*		Ἀείζωον	Aizoum	92
Bupleurum, 2 *sp.*				91
—— fruticosum		Σέσελι αἰθιοπικὸν (Sib.)		91
Hedera, 1 *sp.*		Κιττός, or Κισσός	Edera	94

Modern Greece.	Modern Italy.	Ancient Greek Names.	Ancient Latin Names.	Page
Cornus, 1 *sp.*	2 *sp.*	Κρανία	Cornum	50
Sambucus, 3 *sp.*	3 *sp.*			95
—— nigra		Ἀκτή	Sambucus	
—— ebulus		Χαμαιάκτη	Ebulus, or Chamæacte	
Viburnum, 2 *sp.*	4 *sp.* Tinus		Tinus (Pl.)	119
Lonicera, 5 *sp.*	9 *sp.*	Αἴγιλος ?		96
Ernodea, 1 *sp.*				97
Scabiosa, 1 *sp.*				97
Santolina [a], 1 *sp.*	1 *sp.*			
—— maritima		Γναφάλιον (? Sib.)	Gnaphalium chamæzilon	98
	chamæcyparissus	Πόλιον ?	Polium	
Conyza, 4 *sp.*				98
—— candida		Ἀρκτίον (Dios.?)		
Artemisia, 1 *sp.*	2 *sp.*			98
—— arborescens		Ἀρτεμίσια	Artemisia	
	abrotanum } santonica }	Ἀψίνθιον ·{	pontica (Pl.)? santonica(Pl.)	
Senecio, 1 *sp.*		Ἠριγέρον	Senecio	99
Gnaphalium, 2 *sp.*	1 *sp.*			100
—— stœchas		Ἑλιχρύσον, or Χρυσάνθεμον	Helicrysum	
Stæhelina, 2 *sp.*				101
Pteronia, 1 *sp.*		Χαμαιπεύκη (Dios.)		101
Centaurea, 1 *sp.*				101
Cineraria, 1 *sp.*	1 *sp.*			102
Erica, 6 *sp.*	6 *sp.*	Ἐρείκη	Erice	103
—— arborea		Μυρίκη	Myrica	
	Calluna, 1 *sp.* Arctostaphylus, 1 *sp.*			
Arbutus, 2 *sp.*	3 *sp.*			50
—— unedo		Κόμαρος	Arbutus, or unedo	
—— Andrachne		Ἀνδράχνη	Andrachne	
Rhododendron	Rhododendron, 2 *sp.*			104
Vaccinium, 1 *sp.*		Ἄμπελος τῆς Ἰδῆς		105. & Rom. Husb., 239

* N.B. In p. 98, *for* "Several species of this evergreen under-shrub," *read* "Five species of this genus, but all herbaceous."

Modern Greece.	Modern Italy.	Ancient Greek Names.	Ancient Latin Names.	Page
Olea, 1 *sp.*	2 *sp.*	Ἐλαία	Olea	Rom. Husb., 165
Phillyrea, 1 *sp.*	4 *sp.*	Φιλλυρία		106
Jasminum, 1 *sp.*				106
Nerium, 1 *sp.*	1 *sp.*	Νήριον	Rhododaphne	104
Periploca, 1 *sp.*				109
Convolvulus, 3 *sp.*	1 *sp.*	Σμῖλαξ λεῖα	Convolvulus?	106
Lithospermum, 2 *sp.*		Λιθοσπέρμον?	Lithospermum?	107
Onosma, 1 *sp.*				107
Solanum, 1 *sp.*	1 *sp.*	Στρύχνος	Strychnon	109
Lycium, 2 *sp.*	2 *sp.*			112
—— europeum		Ῥάμνος (Sib.)		
Verbascum, 1 *sp.*		Φλόμος	Verbascum	112
Teucrium, 9 *sp.*				108
—— polium		Πολίον	Polium	
Satureia, 4 *sp.*	1 *sp.*			114
Thymbra, 1 *sp.* (Satureia thymbra)	1 *sp.*	Θύμβρα	Thymbra	
Lavandula, 3 *sp.*	1 *sp.*			114
—— stœchas		Στοίχας	Stœchas	
Origanum, 2 *sp.*		Ὀρίγανον	Origanum	108
—— dictamnus		Δίκταμνος	Dictamnus	
Thymus, 2 *sp.*		Θύμον	Thymus	115
—— zygis		Ζύγις (Dios.)		
Acynos, 1 *sp.* (Thymus acynos)				116
—— graveolens		Τραγορίγανον		
Stachys, 2 *sp.*		Στάχυς?	Stachys?	116
Phlomis, 1 *sp.*	1 *sp.*	Φλόμος ἀγρία	Phlomis	113
Molucella, 1 *sp.*				116
Prasium, 1 *sp.*				116
Salvia, 3 *sp.*	1 *sp.*			115
—— officinalis		Ἐλελίσφακον	Salvia	
Rosmarinus, 1 *sp.*	1 *sp.*	Λιβανώτις	Libanotis, or Rosmarinus	113
Salicornia, 1 *sp.*				118
Salsola, 1 *sp.*				118
Polygonum, 2 *sp.*				119
Vitex, 1 *sp.*	1 *sp.*	Ἄγνος, or Λύγος	Agnus castus	117
Atriplex, 4 *sp.*	2 *sp.*			
—— halimus		ἅλιμος (Dios.)	Alimon (Pl.)	118
Daphne, 13 *sp.*	4 *sp.*	Κνέωρος	Daphnoides	119
—— Tartonraira		Κνέωρος ὁ λευκός		
—— Gnidium		Θυμέλαια (Dios.)	Casia herba (Virg.)	
—— jasminia (Sib.)	oleoides (L.)	Χαμαίλεα	Chamælea	

Modern Greece.	Modern Italy.	Ancient Greek Names.	Ancient Latin Names.	Page
Passerina, 1 *sp.*	. . .	Κνεώρον μέλαν	. . . .	121
Laurus, 1 *sp.*	1 *sp.*	Δάφνη . . .	Laurus . .	119
Osyris, 1 *sp.*	1 *sp.*	"Οσυρις . .	Osyris . .	123
Elæagnus, 1 *sp.*	1 *sp.*	Ἐλαίααἰθιόπικη ? (Sibth.)	. . . .	123
	Hippophae 1 *sp.*			
Aristolochia, 1 *sp.*	1 *sp.*	Ἀριστολόχια .	Aristolochia .	**124**
Euphorbia, 3 *sp.*	1 *sp.*	Τιθυμάλλος (Th.) Τιθυμάλος (Dios.)	Euphorbia,Ti- thymalus .	124
—— dendroides .	1 *sp.*	Τιθυμάλλος δενδρῶδης .	Euphorbia dendroides	
Buxus, 1 *sp.*	2 *sp.*	Πυξός . . .	Buxus . .	125
Ficus, 1 *sp.*	2 *sp.*	. . . .	. . . .	2
—— carica	. . .	Συκῆ . . .	Ficus . . .	
	Sycamorus [b]	Συκαμίνος (ἐν Αἰγυπτῶ) (Th.) Συκόμορον (Dios.)	Ficus Egyptia	
Morus	3 *sp.* nigra .	Συκαμίνος(Th.)	Morus . .	2
Juglans, 1 *sp.*	2 *sp.*			
—— regia .	. . . .	Κάρυον Βασι- λικὸν(Dios.)	Nux Juglans	6
Ulmus, 2 *sp.*	1 *sp.*	Πτελέα . .	Ulmus . .	53
Celtis, 1 *sp.*	1 *sp.*	Λωτός (Dios.)	. . . .	54
Alnus, 1 *sp.*	1 *sp.*	Κλήθρος . .	Alnus . .	55
Carpinus, 2 *sp.*	3 *sp.*	. . . .	. . . .	60
—— betulus .	. . .	. . . .	Carpinus	
—— ostrya (Sib.) (Ostrya vulgaris, Dec.)	. . .	"Οστρυς . .	Ostrya	
Corylus, 2 *sp.*	3 *sp.*	. . . .	. . . .	6
—— avellana	. . .	Καρύα ἡρα- κλεωτικὴ(Th.) Κάρυον ποντι- κὸν (Dios?)	Nux abellina, or pontica	

[b] In my first Lecture, p. 2, I have given for the Sycamore Fig the Latin synonyme of *Sycamorus;* and for the Mulberry that of *Morus*, omitting the corresponding Greek synonymes, viz. Συκαμίνος Αἰγυπτία for the former, and Συκαμίνος alone for the latter. That Theophrastus employed the terms in this sense was pointed out by Bodæus in his notes on Theophrastus, and is confirmed by Fraas (Synopsis Plant. Fl. Cl.) from various considerations.

Dioscorides uses the term μορία and συκαμηνέα for the Black Mulberry, and that of Συκαμίνος (ἐν Αἰγυπτῶ), or simply τὸ Συκόμορον, for the Sycamore Fig.

Modern Greece.	Modern Italy.	Ancient Greek Names.	Ancient Latin Names.	Page
Salix	10 *sp.* . .	'Iτέα . . .	Salix . . .	55
Populus, 3 *sp.* . .	3 *sp.*			
—— alba . . .		'Αχερωίς, or Λεύκη .	Populus alba	55
—— nigra . .		Αἴγειρος . .	—— nigra	
—— tremula				
Fagus, 1 *sp.* . .	1 *sp.*			
—— sylvatica .		'Οξύη . .	Fagus . .	7
Castanea, 1 *sp.* . .	1 *sp.*			
—— vesca . .		Κάστανον . .	Castanus . .	7
Quercus, 9 *sp.* . .	10 *sp.* . .	. . . .	. . . .	10
—— robur . .		Δρῦς . .	Robur	
—— esculus . .		Φηγός . .	Esculus	
—— ilex . . .		Πρῖνος . .	Ilex	
—— suber . .		Φελλός . .	Suber	
—— cerris or Tournefortii . .		Πλατύφυλλος .	Latifolia	
—— toza (dwarf variety of cerris) . .		'Ημερίς . .	Hemeris	
—— pseudo-suber .		'Αλίφλοιος .	Haliphlœos	
—— ægilops . .		Αἰγίλωψ . .	Ægilops	
—— coccifera .		Πρινος . .	Ilex (Pl.)	
	Betula, 1 *sp.*	. . . .	Betulla . .	59
Platanus, 1 *sp.* .		Πλάτανος .	Platanus . .	61
Taxus, 1 *sp.* . .		Μίλυς . .	Taxus . .	42
Ephedra, 1 *sp.* . .	1 *sp.* . .	Τράγος (Dios.)	Ephedra . .	126
Pinus, 4 *sp.* . .		. . . .	. . . .	18
—— halepensis, or maritima . .		Πεύκη . .	Pinaster	
——	Pinaster .	Πεύκη παραλίας	Tibulus	
	Mugho . .	. . . .	Tœda	
—— laricio, or sylvestris . . .		Πεύκη 'Ιδαῖα .	Picca	
—— pinea . .		Πίτυς . .	Pinus	
	Cembra . .	. . . .	Strobus?	
—— excelsa[c]				
Abies pectinata . (Pinus picea, Sib.)		'Ελάτη . .	Abies . .	26
	excelsa . .	. . . .	—— gallica .	22
	Larix . .	. . . .	Larix	
	CedrusLibani	Κέδρος . .	Cedrus . .	33
	Thuya articulata .	. . . .	Citrus . .	41
Juniperus, 6 *sp.* . .		. . . .	. . . .	38
—— communis .	4 *sp.* . .	'Αρκεύθος μίκρα		
—— oxycedrus .		Κέδρος μίκρα		

[c] Lately added to the list by Grisebach, see p. 32.

Modern Greece.	Modern Italy.	Ancient Greek Names.	Ancient Latin Names.	Page
Juniperus phœnicia .	. . . . .	Βράθυς ἕτερον		
—— sabina . . .	. . . . .	Βράθυς		
—— excelsa . . .	. . . . .	Κέδρος ? . .	Cedrus ?	
Cupressus, 1 *sp.* . .	2 *sp.*			
—— sempervirens .	. . . .	Κυπάριττος (Th.) . .	Cypressus .	42
Asparagus, 3 *sp.* . .	. . . . .	. . . . . .	. . . . .	127
—— acutifolius . .	. . . . .	Ἀσπάραγος .	Corruda	
Aloe, 1 *sp.* . . .	. . . . .	Ἀλόη . . .	Aloe . . .	127
Ruscus, 3 *sp.* . . .	3 *sp.* . .	. . . . .	. . . . .	122
—— hypoglossum .	. . . .	Ὑπόγλωσσον (Dios.) . . Δάφνη ἀλέξαν- δρεια (Dios.)	Laurus Taxa, Radix Idæa (Pl.)	
—— hypophyllum .	. . . . .	Χαμαιδάφνη .	Laurus Alex- andrina (Pl.)	
—— aculeatus . .	. . . . .	Μυρσίνη ἀγρία		
Smilax, 1 *sp.* . .	2 *sp.* . .	Σμῖλαξ . .	Smilax	

𝔓𝔯𝔦𝔫𝔱𝔢𝔡 𝔟𝔶 𝔐𝔢𝔰𝔰𝔯𝔰. 𝔓𝔞𝔯𝔨𝔢𝔯, ℭ𝔬𝔯𝔫𝔪𝔞𝔯𝔨𝔢𝔱, 𝔒𝔵𝔣𝔬𝔯𝔡.

CORRECTIONS AND ADDITIONS.

Page 9. *Quercus pubescens.* According to Dr. Alexander Prior, this is not a variety of *Q. robur*, but a distinct species.

Page 10. I have omitted in my list of oaks *Quercus occidentalis* of Gay.

Page 13. *Ilex aquifolium* is now removed from the *Rhamneæ*, and placed by A. Brongriart in a family of its own.

Page 28. A similar tendency to put forth fresh branches when cut down, is noticed in the "Gardener's Chronicle" for August 17, 1861, by Dr. Seemann, as existing in a species of Fir which was found composing a forest in Arcadia, one league and a-half from Tripolitza, called by Heldreich *Abies Reginæ Amaliæ*. It is pronounced by Murray[a] to be the same species as *Abies Apollinis* of Link, which latter was omitted in my work, as having been identified by Endlicher with *P. pectinata*, although Goudon regards it the same as *cephalonica*.

Page 34. The anecdote related with regard to the first introduction of the Cedar of Lebanon into France, was extracted, without enquiry on my part, from an article "On Coniferous Trees," which appeared in the No. of the "Edinburgh Review" for October, 1864.

It turns out, however, to be a tissue of errors, occasioned by blending into one, two stories utterly unconnected, as I might have learnt before printing my Essay, had the remarks of Dr. Asa Gray, inserted in "Silliman's Journal" for March, 1864, fallen in my way when they first appeared.

It is indeed true, that the first seedling of the Cedar of Lebanon which ever reached France, was brought there by Bernard de Jussieu in his hat, just as is repre-

[a] Hort. Soc. Proc., 1863.

sented in an engraving to Cap's popular work, entitled *Le Musée d'Histoire Naturelle*, 1854, p. 15. But the plant in question was brought, not from the Holy Land, which indeed Jussieu had never visited, but from England.

It is also true, that the zeal and self-devotion which the Reviewer erroneously ascribes to Bernard de Jussieu, in preserving the Cedar of Lebanon, was really evinced with regard to another plant, the Coffee, by the individual to whose care it had been confided.

It appears, that when the French Government wished to introduce the cultivation of Coffee into their West Indian Colonies, they despatched a vessel laden with a few of these plants to Martinique. In the course of a voyage unusually protracted through contrary winds, the crew were all placed upon short allowance, and the Coffee plants in general perished of drought. One, however, was kept alive by the captain, who divided with it the scanty portion of water that fell to his share, and this solitary specimen became the parent of all those now found in the Antilles.

The name of the captain, M. Declieux, has been perpetuated by the name *Declieuxia*, given to a genus allied to the Coffee, in remembrance of his services.

By thus jumbling together these two stories, the Reviewer would seem to have concocted the anecdote, which has been unwittingly transferred to the pages of my Essay.

Page 40. Dr. James Mitchell has published in a distinct pamphlet, a fuller account of the Citrus wood of the ancients, than is given in my Essay.

Page 50. *Cornus mascula* is not the Cornel of English Botanists, which is *Cornus sanguinea*.

Page 87. A writer in the Ann. of Nat. Hist., suggests that the Lotus of Homer may have been *Nitraria tridentata*, a plant common in Barbary.

𝔓ublications by the 𝔄uthor.

A DESCRIPTION OF ACTIVE AND EXTINCT VOLCANOS, OF EARTHQUAKES, AND OF THERMAL SPRINGS; With Remarks on their Causes, Products, and Influence on the Condition of the Globe. Second Edition, greatly enlarged. With Twelve Maps and Plates. TAYLOR and FRANCIS, London, 1848. 8vo.

Price with Supplement, 1858, £1 1s.

BRIEF REMARKS ON THE CORRELATION OF THE NA-TURAL SCIENCES. Drawn up with reference to the Scheme for the Extension and the Better Management of the Studies of the University. J. VINCENT, Oxford, 1848. 8vo. 1s.

A POPULAR GUIDE TO THE BOTANIC GARDEN OF OXFORD, AND TO THE FIELDING HERBARIUM ANNEXED TO IT. Second Edition. *Sold only at the Garden.* 16mo. 6d.

AN INTRODUCTION TO THE ATOMIC THEORY. Second Edition, greatly enlarged, 1852. Printed at the OXFORD UNIVERSITY PRESS. 12mo. 6s.

CAN PHYSICAL SCIENCE OBTAIN A HOME IN AN ENGLISH UNIVERSITY? An Inquiry suggested by some Remarks contained in a late number of the "Quarterly Review." J. VINCENT, Oxford, 1853. 8vo. 1s.

LECTURES ON ROMAN HUSBANDRY, delivered before the University of Oxford; comprehending an Account of the System of Agriculture, the Treatment of Domestic Animals, the Horticulture, &c., pursued in Ancient Times. Oxford and London: J. H. and J. PARKER. 8vo. 6s.

CLIMATE: AN INQUIRY INTO THE CAUSES OF ITS DIFFERENCES, AND INTO ITS INFLUENCE ON VEGETABLE LIFE. Comprising the substance of Four Lectures delivered before the Natural History Society, at the Museum, Torquay, in February, 1863. Oxford and London: J. H. and J. PARKER. 8vo. 4s.

www.ingramcontent.com/pod-product-compliance
Lightning Source LLC
Chambersburg PA
CBHW020546270326
41927CB00006B/734